PLANNING SCENERY
for Your Model Railroad

How to use nature for modeling realism

Tony Koester

KALMBACH
BOOKS

About the author

Planning Scenery for Your Model Railroad is Tony Koester's fifth book focusing on his life-long hobby of model railroading. Tony is the editor of *Model Railroad Planning*, a special annual issue of *Model Railroader* Magazine, as well as a contributing editor to MR. He writes MR's popular Trains of Thought column and has written numerous feature articles for MR. He also served as the editor of *Railroad Model Craftsman* magazine until 1981.

Tony spent a quarter of a century designing, building, and with his friends operating a freelanced coal-hauling HO railroad, the Allegheny Midland (the Midland Road), numerous photos of which appear in this book. More recently, he has been hard at work on a multi-deck HO layout that accurately depicts the Nickel Plate Road's St. Louis Division as it appeared in 1954.

Acknowledgements

Among those who made special efforts to provide information, photos, and content suggestions are Chuck Bohi, Jim Boyd, Eric Brooman, Jack Burgess, Mike Confalone, Mike Danneman, Paul Dolkos, Doug Gurin, Jeff Halloin, Tom Johnson, Bob Liberman, Allen McClelland, Bill and Mary Miller, Lance Mindheim, Jack Ozanich, Michael Pennie, Mark Plank, Kirk Reddie of *N Scale Railroading*, C.J. Riley, Paul Scoles, Jim Six, Doug Tagsold, and Jeff Wilson. Jeff and his Kalmbach Books boss Mark Thompson also guided this book through Kalmbach's meticulous production process. Special thanks go to long-time friend and retired geologist Bill Metzger, who I have driven to distraction over the years with incessant questions (and imperfect distillations of his answers) about geology.
Tony Koester
Newton, N.J.
April 2007

On the cover

Clockwise from top: A pumpkin-littered porch and covered bridge bracket a Green Mountain freight heading from Bellows Falls toward Rutland, Vt., on a pleasant autumn afternoon; the author dry-brushes white-water streaks on Coal Fork; an Allegheny Midland mine shifter crosses Coal Fork as it works upgrade toward Low Gap, W.Va. Photos by the author.

Unless otherwise credited, all photography in this book is by the author.

Printed in USA

11 10 09 08 07   1 2 3 4 5

Visit our Web site at www.KalmbachBooks.com
Secure online ordering available

Publisher's Cataloging-in-Publication Data
(Prepared by the Donohue Group, Inc.)

Koester, Tony.
    Planning scenery for your model railroad : how to use nature for modeling realism / Tony Koester.

    p. : ill. ; cm.
    ISBN: 978-0-89024-657-3

1. Railroads—Models. 2. Railroads—Models—Design and Construction—Handbooks, manuals, etc. I. Title

TF197 .K645 2007
625.1/9

Contents

INTRODUCTION

Understanding "scenery"

"Scenery" has a big job to do on a model railroad, including hiding the mechanical underpinnings, establishing a context of place and time, suggesting a reason for being, and convincing the eye that it's seeing something much larger and of greater extent than is actually there. This snow scene Mike Danneman modeled on his spectacular N scale Denver & Rio Grande Western (also see page 74) makes the point.

A model railroad, like a book or play or even a PowerPoint presentation, is at its best when it tells a coherent and compelling story without anachronisms or other evidence of visual or functional disharmony. Moreover, most model railroaders, including freelancers, would not think of scratchbuilding, say, a freight car or a depot without detailed scale drawings and photos. Yet too often we blissfully fill an entire basement or spare room with scenery without equally detailed knowledge of its "prototype." My goal in writing this book is to provide real-life examples of the types of scenic features from rock formations to farmers' crops that typify what you would expect to encounter along railroad tracks in North America.

Looking to the prototype

As its title implies, this book is more about what to do than how to do it, although there are modeling tips and techniques aplenty between these covers. It's intended to complement previous Kalmbach books by scenery gurus Dave Frary (*Realistic Model Railroad Scenery*) and Lou Sassi (*How to Build and Detail Model Railroad Scenes*). They used models to illustrate their points, but here I will more often employ prototype photos so that you can use reality, not (for example) my own imperfect interpretations thereof, as benchmarks. Moreover, I have enlisted the support of a number of outstanding practitioners of the art of building model railroad scenery to literally and figuratively expand our horizons.

It's a lot easier to do anything when you have some sort of benchmark to measure against. This comes almost automatically when you're modeling a specific part of a favorite prototype (full-size) railroad in a narrowly framed time span. You know from the outset what the finished scene, structure, locomotive, or car is supposed to look like. Virtually anyone can readily offer an opinion as to whether you have succeeded.

As I discussed in two previous Kalmbach books, *Realistic Model Railroad Design* and *Realistic Model Railroad Building Blocks*, it's also easier for freelancers when they choose a base prototype or region upon which to pattern their modeling efforts. This is the idea behind the concept of Layout Design Elements (LDEs), the theme of the building-blocks book. The idea is to choose a suitable prototype town, junction, yard, industry, engine terminal, or signature scene and then simplify it a bit to fit your available space without foregoing the visual and operational characteristics of the original.

This advice applies to scenery as well as to track, structures, and rolling stock. Making rocks by smearing a thick coat of plaster on a steep slope and carving willy-nilly may satisfy the artist in our souls, but it's unlikely to appease the civil engineer inherent in anyone who is fascinated by the machinery of railroading. The latter

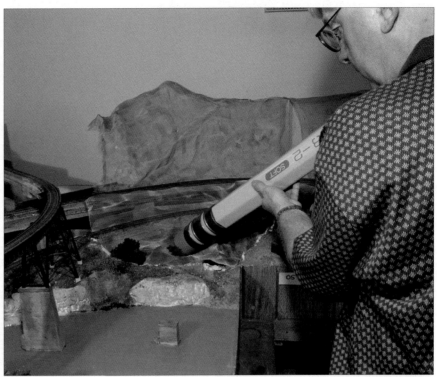

Judy Koester

It's a lot easier to plan plausible scenery from the outset by referring to real-world examples than to have to "erase" it and start all over again.

component of the inner voice that guides us in our modeling endeavors will notice something is amiss. (If not, knowledgeable friends probably will.)

Depending on our powers of observation and training, we may or may not be able to figure out what's wrong. That's where this book comes in. If all of us were talented artists, we could readily see things as they really are and interpret them in miniature for our model railroads. For those who lack such skills and training, some sort of guidelines should help us along. For example, if we know about the major rock types and formations in various regions of the continent, we can make more informed choices of rock molds and castings and coloration. If we know what minerals and crops and industries are commonly found where, we can more readily make informed decisions about what our railroads might do for a living.

One of the railroads' most prolific sources of tonnage and revenue, for example, is coal. Rather than devoting a major part of this book to that intriguing topic, I covered coal railroading in another Kalmbach book,

A Model Railroader's Guide to Coal Railroading. But there are other natural resources that play a role in keeping balance sheets in the black, and we'll touch on them here.

Among the topics we'll visit are modeling various seasons, including autumn and winter; various ways to model water (one size does not fit all); using "texture" to build in a sense of yesterday and the day before; ensuring that your railroad, whether freelanced or prototype-based, reflects its roots, both geographically and chronologically; and learning how to read and use topographic maps.

Last, let me encourage you to follow some age-old advice: If at first you don't succeed, try, try again. As the accompanying photo suggests, although it's better to do it right the first time, "erasing" a false start or two may not be fun, but it can be very satisfying. As your skills improve, your tastes change, or new information, materials, and techniques become available, take another look to see whether you can apply the information in this book to do an even better job the second time around.

CHAPTER ONE

Signature scenes

The heritage of this scene on Paul Dolkos's HO railroad is apparent when it's compared to Philip R. Hastings' prototype photo (page 7) of Barre & Chelsea mixed train No. 3 crossing over the Boston & Maine in front of a Canadian Pacific freight at Wells River, Vt. The steamer was scrapped soon after this July 1946 photo; Paul has modeled the GE 70-ton replacement.

A visitor's first look at your railroad will be very much like an introduction to another person: It will likely form a lasting impression. If you can give visitors visual clues as to your railroad's location, era, and purpose right up front, you can spend more time explaining other aspects of your railroad that you're especially proud of. It's therefore in your best interests to greet visitors, whether they see your railroad in person or through photographic images, with a "signature" scene (**1-1** through **1-6**) that clearly tells them a key part of the story you want them to hear.

Philip R. Hastings

Nailing it down

As well-respected model railroad designer John Armstrong once remarked, it's a lot easier to model a specific prototype than to do a convincing job of freelancing. The reason is apparent in the two photos that grace the top of this spread: If something actually exists or existed, the main task is to replicate in miniature its key features at some point in time. There's little need to "invent" anything.

Given today's cornucopia of well-detailed locomotives, rolling stock, structures, track, and building materials, faithfully modeling a prototype isn't quite the chore it used to be. That said, the bar has been raised; today's better models are the result of more knowledgeable, and hence demanding, modelers and responsive manufacturers. So doing a good job of modeling a specific prototype, or a freelanced amalgama-

tion of several similar prototypes, still warrants doing considerable homework. The good news: That's easily half the fun.

The prototype-based freelancer faces all of those challenges plus others: He or she has to create something that looks familiar and plausible, even though we've never seen it before (1-4). Basing a freelanced model railroad on familiar yet slightly altered slices of reality may therefore pay big dividends.

(A caveat: Freelancers may want to read chapter 7 of *Realistic Model Railroad Building Blocks* for a discussion about when to crib and when that may be confusing to the viewer.)

In either case, the closer one's modeling efforts are to some well-known geographic locale, the easier it is for even a casual visitor to relate the model to full-size railroading in that region. That is not to say that one must model

famous scenes, such as Horseshoe Curve, Sherman Hill, or Starrucca Viaduct, for a model railroad to be seen as realistic.

Indeed, it's usually more effective to model something far more typical – the railroad side of a small town or city that looks like something out of an old Jimmy Stewart movie will do quite nicely. In that case, the signature scene may include no signature structures, yet overall it will convey the desired sense of time and place.

For example, the glaciated, hence rounded, ridges of areas north of the Ohio River contrast sharply with the more pointed peaks of the southern Appalachians, let alone the Western ranges (chapter 4). Old rivers and creeks of the East and Midwest meander (1-13), as streams do over the eons; younger Rocky Mountain streams race through V-shaped valleys toward the

ocean – except in the northern Rockies and Sierras, where repeated encounters with flowing ice sheets have hewn out U-shaped valleys. And the palm trees of southern climes (see "The Sunshine State" on page 10) are unlikely to be confused with the pines of the Carolinas or deciduous trees of northern Appalachia.

"Events"

When you have a clear view of a train moving from one end of your railroad to the other, that constitutes one visual "event." You can take in the action with one look. It follows that if we erect something that temporarily blocks our view of the train, its passage will be more interesting – it becomes two or more events.

Such view blocks can be an overhead bridge, a tall foreground structure (**1-7**), a hill or mountainside, a tunnel, or even a stand of trees. The idea is to have the train repeatedly disappear

from sight, then reappear again, as it traverses a major section of your railroad.

I complemented this idea on the Allegheny Midland with "lobes" (mini peninsulas) between towns (**1-8**). The towns were located more closely together than I would have liked, but the view-blocking lobes, often sporting ridges with tunnels, made it harder to see from one town to the other at a glance.

The town of Thurmond, W.Va. (left), is jammed between the New River and an adjacent ridge without room for a main street between the tracks and the storefronts. The C&O (CSX today) had an engine terminal for mine shifters here. Thurmond served as the inspiration for a similar scene at Sunrise, Va., on the author's Allegheny Midland.

1-4

A towering white church steeple jutting above deciduous trees isn't unique to New England, but it helps form a signature scene nonetheless. Boston & Maine Mogul 1470 (opposite) chuffs through picturesque Milford, N.H., in this 1949 portrait. Paul Dolkos included similar churches on his HO B&M layout (above). The one up front is a Campbell Scale Models kit; the other is a photo pasted on the backdrop.

1-3

"Signature" doesn't mean "spectacular"

A signature scene does not have to be a "trophy" scene. It need not be spectacular or even especially noteworthy. Its mission is to identify the locale being modeled (**1-9** to **1-12**); if it provides clues as to the era, so much the better.

When picking signature scenes and structures, look for helpful place names. Western Indiana Gravel, for example, ran a number of gravel pits in the Hoosier state, and the company name often appeared on entrance signs and office buildings. Modeling these structures and signs would help viewers locate even a freelanced model railroad.

Considerations of scale

The debate about which scale is "best" will rage on as long as there are model railroads and model railroaders, but the fact is that each scale has useful advantages and annoying disadvantages. The key is to match your modeling goals with the pluses and minuses of each scale, and then choose a path of least resistance or, put more positively, greatest opportunity.

Consider, for example, Lance Mindheim's creek scene on his N scale depiction of Indiana's Monon Railroad (**1-13**). Clearly, it's an old creek, as it meanders back and forth like a laid-flat sine wave as it negotiates and renegotiates a path through the pasture.

From track center line to viewing-aisle fascia, this scene measures about 21", or 280 scale feet (one actual foot equals 160 scale feet in N scale). That's more than enough distance to show the train running along a tree line that defines a low ridge on the far side of a sizable pasture. Yet it doesn't violate good layout-design protocol – at most, a 30" reach-in distance – as Lance can still reach in to clean or repair track or rerail an errant car.

Now let's consider modeling that same scene in O scale. At 1:48, that 280-scale-foot reach-in distance translates to 70", just under 6 actual feet, which is obviously a problem. Even in HO, the reach-in distance is 39", a bit of a stretch.

The solution is to adhere to the 30" reach-in maximum and model less of the foreground. You'll still have enough of the scene to be convincing, if not quite as dramatic, in HO, and in O you'd trade the over-the-fields scene for a more dynamic view of big trains running close at hand.

Harsh New England winters caused some freight houses to be built over the house track, including a structure at Milford, N.H., that Paul Dolkos used as a pattern for this HO model at Woodsriver (left). Such structures were not limited to New England, however; the example (right) was on the Monon at Gosport, Ind.

1-5

1-6

Lance Mindheim modeled an HO scale industrial spur based on a prototype in Miami. The pastel-colored stucco structure, sabal palmetto tree, cabbage palms, and Spanish-language billboard all reinforce the identity of the region and area modeled.

The Sunshine State

When thinking of signature scenes, the state of Florida may not come readily to mind. But if you get it wrong, the myriad people who have visited the Sunshine State will spot any missteps in short order. Like other regions of the continent, it has its own distinctive, and highly modelable, look.

Case in point: My HO scale CSX East Rail switching layout (**1-6**) depicts a specific industrial park spur in Miami. The era is present day. In putting together a scene like this, I incorporate elements that will tell the viewer both where the scene is located and during what time period. Structures, vegetation, and rolling stock are all helpful tools that tell a person what time and place they are looking at.

For example, structures in this region often share a common construction method: cinder block covered with stucco. Colors are bright pastels faded by long strings of sunny days. The structure in this photo represents Guixens Food Group, a food distributor that handles bagged rice as well as other food products. Notice the stucco finish, faded pastel paint, and bags of rice inside the loading dock door.

Vegetation can also be a helpful tool in telling people where a scene is located. The sabal palmetto (the tall palm on left side of structure, available in kit form from Hart of the South, www.hartofthesouth.com) is one of the more common trees in the area I'm modeling, as are the small cabbage palms (under the crossbucks) that tend to sprout like weeds.

Even by U.S. standards, Miami is a melting pot of all nationalities and their associated languages. The billboard on top of the structure was copied from the prototype and provides an opportunity on several fronts to nail down both time and place. Certainly the Spanish text is one clue as to location. If you look closely you'll see other hints as to the time frame being modeled. On the billboard frame, note the logo of Clear Channel Communications, a modern era corporation. The main advertiser, H & R Block, is a modern business as well.

– Lance Mindheim

Scenic intensity

Unless you have a lot of spare time or an army of skilled helpers, the size of your layout will have a marked effect on the intensity of each major scene on your railroad. The achievable intensity of scenic detail is a function of the magnitude of the overall project. The modeler who focuses on building a narrow shelf layout across one wall of a spare room can afford to devote much more time to modeling every nut and bolt than can the guy or gal who wants to fill a gymnasium with most of an entire division.

I know of very few fine-scale layouts – Proto:87 (**1-14**) or Proto:48, for example – that fill large basements. Modeling to fine scale (almost exactly to prototype) track and wheel standards is a lot like modeling back in the hobby's formative years when "ready-to-run" meant "toy train set." Everything had to be assembled from craftsman kits or built from scratch.

Today, fine-scale modelers can get some commercial help, but it's not likely to be as close as the nearest hobby shop, model railroad club, or National Model Railroad Association division meet. The rewards are considerable, but you'll spend a lot more time building than operating for quite some time.

Even those who are happy with today's wheel and track standards may enjoy intensely detailing their structures and scenery. This requires more effort, and hence time, so the sheer size of a layout may need to be scaled down to compensate, or at least built in finite stages so that one area can be declared finished before moving on.

And, as Andy Reichert of The Proto:87 Stores (www.proto87.com) points out, modeling an urban scene can be expensive. Coming up with enough pre-built vehicles and streets lined with continuous rows of structures to "do it right" may push the density and hence cost of an urban setting much higher than its rural counterpart. And it's worse at rush hour: "You shouldn't have empty platforms and trains on 20-minute headways, or full platforms served by one train a day," he observed.

Photo labels (1-8): GLADY 52.0, GL, Culvert, BIG SPRINGS JUNCTION 52.5, BJ, LEAD MINE (R&MC staging) 44.5, ELKINS (WM staging), eek, over, Run, Cheat River, 50.4, 1, 6, 1-8

Photo label (1-7): estvaco, P&LE, LEHIGH VALLEY, 4, 1-7

A single scene can be divided into two or more "events" by having a train pass behind a structure, hillside, or stand of trees or through a tunnel. Care must be taken not to block access to busy areas, however.

"Lobes" are mini-peninsulas that serve as view blocks between adjacent towns, thus providing greater visual separation. Two lobes are apparent here below Big Springs Junction and left of Glady, W.Va., on the author's Allegheny Midland.

Urban modeling has an appeal all its own, as Bob Smaus documented in the July 2006 MR and on his Web site: www.bobsgardenpath.com/trains.html.

Floor-to-ceiling scenery

The late John Allen was a master of illusion. In some areas, he extended the scenery on his HO Gorre & Daphetid layout all the way down to the floor, then positioned his camera near floor level and shot up at the trains to give the viewer the sense of a railroad perched high up on the side of a mountain. His railroad wasn't really all that high off the floor, however; the published photos just made it seem that way.

Similarly, Doug Tagsold built the scenery for his On3 Rio Grande layout in a relatively conventional manner except for one area where he wanted to provide viewers with a dramatic view of the Silverton Branch's High Line section above the Animas River. To accomplish this, he looped the railroad

back from the main aisle (1-15). Viewers see this area from the open end of the canyon rather than walking into the scene.

When I visited Doug's layout, I asked him how he got into the scene to clean track, rerail a car, or work on the scenery.

"The floor-to-ceiling scenery is in a 10 x 12 foot area," he explained. "Before constructing any benchwork in the area, I first covered the already carpeted floor with several layers of heavy plastic drop cloth, then laid 2 x 4s on edge on 16" centers and covered them with ¾" plywood, creating a deck 4" above the floor. The benchwork and scenery were then built on top of the deck, so that when the time comes to dismantle the railroad and the deck is removed, the original floor and carpet will be unaffected.

"After trackwork and scenery in the area were finished, I used Enviro-Tex resin to form the river. Waves and rapids were then made using clear and

white silicone caulks. On the two occasions when I had to enter this area, I removed my shoes and carefully walked in. The plywood deck proved strong enough to support my weight without sagging or cracking the Enviro-Tex."

You can see more photos of Doug's handiwork and his 24 x 50-foot track plan in *Great Model Railroads 2007*.

Using model photos to advantage

There are two important scenery lessons inherent in John's and Doug's examples of scenery that extend well below the benchwork. First, we need to consider the space below the benchwork, perhaps all the way down to the floor, if the objective is to create the sense of a deep canyon. Second, we can use photographs of our model railroads to give viewers a more realistic or dramatic perspective of them than they might actually see if they visited the railroad.

On one short scene on the Allegheny Midland, trains popped into view

Coal mines dotted the landscapes of Colorado as well as central Appalachia. Mine structures ranged from wood anthracite breakers, like this O scale model Bill Miller built with coffee stir sticks (top), to modern corrugated-metal prep plants like this example Eric Brooman built for his HO Utah Belt (above). **1-9**

pains not to create curves sharper than that minimum.

But too often we then make our minimum radius our standard radius. Even when an opportunity to create a much broader, more visually appealing "cosmetic" curve presents itself, we often don't sense the opportunity at hand.

I set 42" as the minimum radius on my HO Nickel Plate Road after tests confirmed that 2-8-4s (**1-17**) and full-length passenger cars looked good on such curves. I had a head start here, as Bill Darnaby had already done similar tests with his 4-8-2s, and they also pointed to 42"-radius curves.

But consider this: Although 42"-radius curves are generous for HO, Frank Hodina and I somehow managed to fit them into the same basement where 30" curves had reigned supreme on the Allegheny Midland. Moreover, had I been an O scale modeler, I would have found a way to accommodate 60" or larger curves. The moral of the story is to be careful what you aim at, as you may hit it. And aim high, not low.

I did discover a few opportunities to use much broader curves. The scenery in those locations is rudimentary at this stage, but already I find myself "railfanning" the railroad as Berkshires lean into the superelevated curves (**1-17**). Take a hard look at your track plan or railroad to see whether similar opportunities exist.

Lighting

Layout lighting deserves an entire chapter, if not a book, but it's a story that is still being written and is likely to change dramatically in the next few years as light-emitting diodes (LEDs) come to play a major role in home lighting. For our purposes, let me cite a few statistics and offer an opinion or two based on personal experience and preference.

A brief physics lesson: Lighting from a point source, such as a light bulb, falls off with the square of the distance. Simply put, if you move the object being illuminated (a model or your daily newspaper) from two to four feet away from a light bulb (double the

for a few feet as a sanity check to make sure they were still moving as they spiraled around a turn-back curve. I added a water tank and depot at what became known as Gap Run, Va. (**1-16**). This scene was featured on the cover of *Model Railroader* and, as a result, became a signature scene for my freelanced railroad. The photo imparted a sense of reality to this scene far beyond

its actual compact size, and I never saw Gap Run in quite the same way again.

Cosmetic curves

Floor space is always at a premium, and broad curves can eat up a lot of it. We therefore choose a radius that our longest cars and locomotives can negotiate without undue visual and operational tribulations, and we take

1-10

Signature scenes need not be spectacular; their job is to inform the viewer about the time and locale of a scene on a model railroad. Here the steam locomotive, railroad name on the tender, and the golden summer grasses provide strong clues about California of the 1930s or 1940s on Jack Burgess's fastidiously detailed HO railroad.

distance), only one-fourth as much light will reach the object being illuminated. Do the same thing with a linear source, such as a fluorescent tube, and you lose only half the original light intensity. Moreover, fluorescents are significantly more efficient, so they cost less to operate and put out less heat.

Point sources can create cones of light on a nearby backdrop and pools of light on the railroad. Fluorescent tubes do much better here, but allowing even short gaps between fixtures can create unwanted shadows on the backdrop at regular intervals (**1-18**). The shadows are easier to disguise on a serpentine mountain railroad, but look like fence posts along the backdrop on a linear Midwestern layout like mine.

Allen McClelland uses screw-in fluorescent "bulbs" to illuminate his

Both images: Charles Bohi

Not much doubt about the general area of these two photos: the Canadian prairies. The depot at the end of the main drag is in Gravelbourg, Sask., in 1986; a pair of CN GMD-1s works a row of "prairie sentinels" at Laird, northwest of Saskatoon. Both scenes could be modeled on a relatively narrow shelf without losing any character.

1-11

Jim Boyd

Signature scenes can be modeled literally, as the Red Rocks area of Willsboro Bay on the D&H (left) and in HO on the New England, Berkshire & Western layout at Rensselaer Polytechnic Institute (top right); or figuratively, as the creek-overhanging porches along the N&W in southern West Virginia (middle right) and at South Fork on the author's Allegheny Midland (bottom right).

1-12

1-13

Lance Mindheim

Modeling an expansive scene with the railroad set well back from the aisle, as in this example showing a creek meandering through a field on Lance Mindheim's N scale Monon RR, may prove problematic in larger scales.

1-14
John Wright

John Wright models American prototypes in HO fine scale (Proto:87) in England with spectacular results. Iron City, while generic, offers myriad lessons on how to model an urban area effectively in the steam era when virtually everything was bathed in soot. He varies the era from 1948 (steam) to 1952 (diesel) but notes that a few years later, the incessant drive toward modernization would "sweep so much away." His PRR Federal Street layout was featured in the May 2003 *Model Railroader* and issue no. 155 of *Model Railway Journal* (U.K.).

Mary Miller

Doug Tagsold

The famed Animas River Canyon and Rio Grande (now Durango & Silverton) High Line north of Durango, Colo. – shown at left from the valley floor – are featured in an alcove of Doug Tagsold's spectacular On3 railroad (right).

1-15

Gap Run, Va., on the author's HO Allegheny Midland, was actually just a short section of visible main line between two tunnel portals that assured crews their trains were still moving. Only a depot and water tower marked the location. But this photograph gave it a new dimension; after it was published in MR, Gap Run became "real."

new single-deck Virginian & Ohio, and they seem to work very well indeed. As the lamp-to-layout distance decreases, however, as it is bound to do on a multi-deck or narrow shelf-type layout, you may start to see cones of light on the backdrop. Run a few tests to check for these effects before making a commitment. And if you're not well versed in wiring 120-volt AC circuits, hire an electrician!

The color of fluorescent lighting to use is highly subjective. GE Chroma 50 tubes or their equivalent, used in rooms where color-printing corrections are made, provide an almost pure-white color. They look very cold to my eye, however, so I stick with less expensive and easier-to-find cool whites (especially in the smaller sizes needed for under-cabinet fixtures). Similarly, warm

white tubes are too red for my tastes, causing pinkish smears on the blue sky backdrop, but they may look fine if you're modeling more arid regions.

Don't even think of using Christmas-tree bulbs or rope lights. The former create dim red pools of light; the latter barely provide enough light for you to find them in the dark. A friend planned to run three parallel strings of white rope lights to illuminate the lower deck of his layout; he now has them for sale, cheap.

Blue rope lights, however, do a very nice job of simulating a moonlit night. I am therefore experimenting with a single string mounted behind the valance, and the tests look very promising. However, the railroad wasn't built to look pretty, but rather to re-create the operating patterns and practices of

its prototype in the steam-to-diesel transition era, and busy operators may find night lighting more of a distraction than it's worth.

Dan Zugelter built his C&O layout in a room with the wall and ceiling corners coved (rounded). The Appalachian ridge line around all four walls is all at roughly the same elevation (for reasons we'll discuss in chapter 4). The tops of the ridges stand out about 1.5" from the sky backdrop, and below the tops of the ridges he has connected separate Christmas-tree light strings with white, yellow, orange, and red bulbs. As the ceiling spotlights dim, the white bulbs come on, creating the impression of the early stages of a sunset. Then the color reddens as the other hues light up, and finally all of the light fades into a dark night. Spectacular!

The author set a 42" minimum radius after tests confirmed that a Nickel Plate Road Berkshire (2-8-4) could traverse these curves without excessive cab or pilot overhang (left). In a few places, much broader "cosmetic" superelevated curves were used for an even more prototypical appearance (right). **1-17**

One last point about light intensity: As you age, your night vision deteriorates dramatically. What seems bright enough when you build your railroad at age 30 or 40 may seem rather dim at 60 or 70. Plan ahead!

Information sources

Throughout this book, I've listed a number of Web sites that offer information of potential value to modelers. The variety of information available through such Web sites is almost limitless. For example, one that lists penny postcards state-by-state and then by county (www.rootsweb.com/~usgenweb/special/ppcs/ppcs.html) may provide glimpses of the way things appeared in the region and era you're modeling.

However, be aware that Web surfing can consume huge quantities of time and become a hobby in itself, so keep your priorities in mind.

A matter of balance

As you're planning your next layout and reviewing your options, be sure to match your scenic goals to the layout's overall size and operational objectives. If you can live with minimal scenery and commercial track, you are more likely to be able to manage a large layout. If you want every scene to have as much detail as the typical craftsman kit, think smaller – or enlist the help of a lot of talented friends.

1-18

The author found that spacing under-cabinet fluorescent fixtures even a few inches apart created a noticeable shadow on the lower-deck backdrop. For single-deck railroads, where there is more distance between the light and the railroad, coil-type fluorescent "bulbs" like those used by Allen McClelland on his new Virginian & Ohio layout (January 2004 *Model Railroader*) work very well.

CHAPTER TWO

Signature structures

Among the better-known and highly admired practitioners of New England railroading in miniature is Paul Dolkos, who has modeled the Wells River, Vt., to Berlin (say BER-lin, not ber-LIN), N.H., line of the Boston & Maine. "I've moved things around to suit my space," Paul reports, "so I'd call it a historical novel as opposed to a history book." He has incorporated actual scenes, structures, and operations from the prototype into his HO layout wherever possible.

Like signature scenes, individual structures play major roles on our model railroads. They help us identify the era and locale of a model scene. Just as mine tipples evoke thoughts of railroading in the Appalachians, Rockies, or even parts of the granger belt, grain elevators – the so-called sentinels of the prairie – bring to mind the Midwest or central Canada. If someone mentions "New England," you can't help but think of low, wooded ridges, white churches with towering steeples (**1-3**), covered bridges (cover and **2-1**), Greek Revival architecture with distinctive trim details (**2-2**), and mill buildings (**2-3**).

That regional look

Buildings with a European flair (2-3) are most common in the East, as that's where most settlers from across the Atlantic first established new roots in North America. Canning plants were typical throughout the Midwest; I'm modeling the large plant owned by Kemp at Frankfort, Ind. (2-4), which conveniently will cover a lengthy stretch of backdrop. Paper mills are found throughout the northern states and Canada, as I discussed in *Realistic Model Railroad Building Blocks*. Produce warehouses populate regions where oranges and apples and potatoes are grown. Tobacco and cotton fields and warehouses permeate the South (see chapter 8).

The key is to identify architectural benchmarks from the region and era you're modeling. Regionalisms, hence billboards and signs, include brands of gasoline, soft drinks, and beer. The local historical society or library may have phone books or city directories for the desired period with listings for local businesses.

And, regardless of the region, you are likely to see more white paint than you may expect (2-5). For brick buildings, you'll often find that the window trim is painted black, white, or green.

Paul Dolkos

Typical New England architectural details such as the cornice returns are evident in these two photos of a general store on Paul Dolkos's HO B&M (top) and a frame building along the Maine Central (above) in 1978. The red paint is a bit unusual; white paint predominates.

2-2

The once-electrified Claremont & Concord ran north down the street toward the mill above the white van (left), then headed west to a paper mill or followed a switchback to go right to downtown. Both Greek Revival- and European-inspired architecture details (right) are evident in the buildings at this intersection.

2-3

Jim Ostler; courtesy Frankfort Public Library

In the steam era, the large canning factory at Frankfort, Ind., was operated by Kemp, as the sign proclaims. The author's HO railroad will feature a full-length model of this plant to help cover a long stretch of backdrop.

In most regions of North America, white paint is clearly favored for commercial and residential structures alike. The store and home in rural Glady, W.Va., along the Western Maryland (left) make the point. Brick structures (right) often have white trim, but green and black are also common.

2-5

Coherent parts of the whole

Some model railroads look more like a display of craftsman-style structure kits than a coherent miniature of any actual place and period. Any of the structures could have been a contest winner; seen as parts of a greater whole, however, they failed to do their jobs.

For those who enjoy building super-detailed structures from scratch or kits, the enjoyment of creating a miniature masterpiece may be the focal point of the hobby. It's like painting a picture in 3-D. But I view each model as part of a whole. Just as running a local freight is little more than an exercise in switching cars unless its work is considered in the context of the through freights that forward those cars to distant climes, I look at scenery and structures as being complementary to the function of the railroad itself.

I have scratchbuilt quite a few structures, kitbashed myriad others, and fashioned countless scale acres of scenery over the years. I enjoy those aspects of the hobby as much as most others.

Roger Thomas built and photographed this impressive – in terms of both size and quality – model of part of an automobile plant based on a Ford plant at Mahwah, N.J. Modeling one large industry instead of an assortment of smaller ones can generate more traffic while making a more lasting impression on viewers.

2-6

(Wiring, not so much.) So my point here is not that realistic operation is the only reason to build a model railroad. But no single aspect of our broad-shouldered hobby should be considered in isolation if we hope to create a plausible model railroad.

To that end, we need to examine structure candidates in the same critical light used to make motive-power and rolling stock choices. No matter how much we may admire that 2-6-6-6 or

2-7

The lengthy brick depot on the Erie at Susquehanna, Pa., stands today as a monument to pre-sleeping-car days when rail passengers spent the night in lineside hotels. This HO model on Harold Werthwein's Erie is a signature structure in every respect.

Illinois Central's *Panama Limited* passes the Briarwood Plantatation's antebellum mansion, a signature structure at Alligator, Miss., on Allen Keller's Bluff City Southern (featured in the November 2002 *Model Railroader*).

Lou Sassi

2-8

4-8-8-4, it will not look at home in Wisconsin. That EMD DDA40X will crowd clearances in New England. And a GE 44-tonner will seem a bit overwhelmed on UP's Sherman Hill.

One big vs. many small

It's often more effective from both a scenic and operational standpoint to model one large industry rather than several small ones. Industries that shipped more than an occasional carload were typically rather massive complexes.

An HO model of an automobile plant built by Roger Thomas (**2-6**) is a good example of an industry that clearly benefits from rail service. In the same space, Roger could have built quite a few smaller industries that might have been appropriate for a small town – a lumberyard, fuel dealer, team track, and grain elevator, for example. But all of these together would not generate nearly the traffic of a single auto plant, nor would they be as likely to leave a lasting impression on the viewer – or be as satisfying to build.

The same reasoning applies to large single structures. A model of a key depot (**2-7**) or a Southern mansion (**2-8**) will draw more comments and more quickly identify the railroad and its location than several lesser structures.

A series of small structures may add up to a noteworthy scene, however. Charcoal kilns and coke ovens were once common in central Appalachia (**2-9**), so they were signature structures on Allen McClelland's original Virginian & Ohio. Mines and stamp mills once dotted the flanks of the rugged Rockies (**1-9**, **2-10**) and hence identify a model railroad's location and era.

Size and proportions

In the days before Walthers produced massive industrial building kits from steel and paper mills to gravel and cement plants, most commercial building kits were sized to fit in relatively small boxes. Stores were often little more than 3 x 5 inches in size. Anything we can do to disguise their original size and proportions will add to their plausibility. This is usually no harder than combining two or more

Allen McClelland

Charcoal kilns (top and above left) and coke ovens (above right) are typical of the Virginias and therefore are signature structures on Allen McClelland's HO scale Virginian & Ohio.

2-9

identical kits to make them longer and/or wider (**2-11**).

But sometimes smaller is better. Placing a smaller-scale structure in the distance or up on a mountain (**9-3**) may lead the eye or camera into seeing more distance than is actually there. Smaller images on backdrops do the same thing.

Ignore the label on the box

When you're looking for structure kitbashing candidates, ignore the label on the box. The soybean plant in 2-12 was hidden in a box Walthers labeled as Valley Cement Co. (I explained how I modeled it in a special 2007 issue of MR, *How to Build Realistic Layouts, Industries You Can Model*.) The Western Maryland-style depot shown (**2-13**) is a kitbash of the Faller farmhouse. Rix billed the two freelanced depots as residential houses; by lowering the roof angle to 30 degrees, they become passable depots or yard offices. The coal tipples came in boxes labeled "Tucson Silver Mine" (Con-Cor) and "Carnegie St. Manufacturing" (City Classics).

I look only at the picture on the box without considering what it is supposed to represent. But even when the box label matches the general description of the structure you need, creativity can pay large dividends. Harry Bilger and Jim Six needed grain elevators for their HO railroads, one set in Idaho, the other in Indiana. They referred to prototype photos and then combined plastic kits into reasonably accurate miniatures (**2-14**).

Stand-ins and mock-ups

You can show what industry a track serves when you don't have time to build each structure by making photocopies of photos or drawings and dry-mounting them on cardstock, then propping them up at trackside. Similarly, easily assembled kits or ready-built models can serve as stand-ins until the true-to-prototype structures can be built.

If you're not sure how structures will fit into a scene, it pays to mock them up. John King built mock-ups of the industrial part of Winchester, Va. (**2-15**; also see *Model Railroad Planning* 2005), to be sure everything would work

2-10

Harry Brunk

Harry Brunk's semi-freelanced Colorado narrow-gauge railroad features models of actual structures that typify the region and era he models – here the Capital Prize Mine near Georgetown.

2-11

Two Smalltown USA (Rix) stores were combined to double the length and width and create a company store for the author's Allegheny Midland.

2-12

The author kitbashed a Walthers Valley Cement kit into a selectively compressed model of the Swift soybean processing plant at Frankfort, Ind., a signature structure on a Midwestern railroad and the largest industry on his new Nickel Plate Road layout.

Ignore the labels on structure kit boxes. The author kitbashed the Western Maryland depot (top) from a Faller farmhouse, the two depots (center) from Rix houses, and the coal tipples from a City Classics factory and Walthers coal mine parts (bottom, in the foreground) and a Con-Cor (Heljan) Tucson Silver Mine. **2-13**

together within his space limitations. Lacking plans, I referred to photos by Gene Huddleston to scratchbuild a Chesapeake & Ohio depot; I assumed battens were spaced a foot apart and doors were 30" wide. Before building the actual model, I made a cardstock mock-up so I could see how various components lined up when compared to prototype photos.

Structure modeling options

If a structure is important to the theme of the railroad, it should be an accurate model of its prototype. For example, the C&O yardmaster's office at Quinnimont, W.Va. (**2-16**), set the tone for all lineside structures on the Allegheny Midland. Unless you're modeling Colorado (**2-17**), the odds are poor that a particular structure will be available as a commercial kit. Moreover, using a popular commercial kit without modifying it invites viewers to comment on how well you did with that kit as opposed to seeing it as a miniature of its prototype.

If I have the information I need about a particular structure – dimensions, perhaps scale drawings, and photos – I usually build the model from scratch. With today's modeling materials, scratchbuilding is like building a kit without having to read the instructions.

If I don't have significant bits of information such as a photo of one wall of the building, I may hold off scratchbuilding it, press ahead but make that wall removable, or combine parts from one or more kits to make a stand-in for the building. I am also more likely to kitbash a structure if I know it won't fit on my railroad if built fully to scale. For example, the silos of the Swift soybean plant (**2-12**) would tower 106 feet, bumping into the overlying second deck, so I was more comfortable kitbashing the structure using 65-foot-tall Walthers silos.

As I discovered on the Allegheny Midland, however, some kitbashed stand-ins grow roots. I never did get around to scratchbuilding several coal tipples after their kitbashed subs turned out well (**2-13**). On a freelanced railroad, no harm done; on a prototype-based railroad, this could become an irritant.

Harry Bilger built an HO model of the Moscow (Idaho) Seed Co. based on photos he had taken in August 1987 (above). He kitbashed it using parts from the Walthers Valley Growers elevator and New River Mining Co. kits plus Evergreen and Model Builders Supply shapes and siding (top right). He drew his own plans from field measurements and Sanborn maps. Jim Six used a similar approach to kitbash a typical Indiana grain elevator (right) using several Walthers and Pikestuff kits and detail parts. By rearranging the main components, an entirely new look was achieved.

2-14

John King used cardstock mock-ups of the entire town of Winchester, Va., on a Baltimore & Ohio branch to ensure everything would fit as planned (left). The author estimated the dimensions of the Chesapeake & Ohio's Sproul, W.Va., depot from photos, made a cardstock mock-up to check proportions, then scratchbuilt the model (right).

2-15

The Chesapeake & Ohio yardmaster's office with a "cabin" atop at Quinnimont, W.Va. (above), was a signature structure that inspired all lineside structures on the author's Allegheny Midland. He therefore measured the building and scratchbuilt an HO model (right). The C&O engines belong to Karen Parker.

2-16

Look familiar? Not every interesting structure in the state of Colorado has been offered as a kit, but many of them have, including this restored brick office building in downtown Salida on the Rio Grande. With a little disguising, it could be used more anonymously anywhere.

2-17

Typicalities

If you think in terms of an individual structure for your layout, you may choose a building that's a bit over the top, something that would stand out at a theme park. If plausibility and realism are important, however, it may be better to think in terms of groupings of structures. Alone, each building may not amount to much, but together they may constitute a believable town, farm, elevator, lumberyard, or whatever.

An orderly row of look-alike company houses (**2-18**) and a company store will convey the look and feel of a company town. In mountain country, where towns and streams fought for space in narrow valleys, the porches may overhang the waterways (**1-12**) and narrow streets (**2-20**). Graph-paper-precise grids of streets embracing the railroad tracks suggest a granger village that may have been built after, or as, the railroad arrived to provide a connection to the outside world. A typical small town's down-by-the-railroad-tracks section may not feature anything of architectural merit, but if its very blandness is modeled convincingly, it will add incredible realism to a model railroad. Just don't forget to use a lot of white paint.

Structure modeling resources

A little homework will uncover books and videos related to your next modeling project. Want to learn more about grain elevators, for example? Lisa Mahar-Keplinger wrote a book called *Grain Elevators*, published by Princeton Architectural Press in 1993. Kalmbach's soft-cover book series covers everything from lineside industries and details to yards and engine terminals.

I often use Google's search engine to find photos, maps, and other material related to modeling projects. The Library of Congress (memory.loc.gov) is a good place to start. Need info on Midwestern barns, for example? Try www.ohiobarns.com/index.html. Grain elevators? Try grainelevatorphotos.com, bnsf.com/markets/agricultural/elevator/index.html, or royalalbertamuseum.ca/vexhibit/elevator/index2.htm, among others.

Company houses, such as these modernized examples at Trammel, Va., near the Clinchfield (top) are signature structures for any Appalachian coal hauler. The author kitbashed these homes (above) from AHM/IHC Speedy Andrews Repair Shop kits; others were made from Grandt Line and Rix houses. **2-18**

2-19

A company store like this distinctive wooden example, shown here in July 1975 on the C&O at Yolyn, W.Va. (above), is another signature structure for a coalfield railroad. Note the typical raised porch. It was previously painted gray.

Mountain railroads and the towns that supported them and the coal industry were typically jammed into narrow valleys with porches overhanging narrow streets in the front and creeks in the back (see 1-12). This scene is near Davy, W.Va., in 1973.

2-20

3-1

Dave Frary

CHAPTER THREE

Texture

Thatcher's Inlet, a tiny HOn2½ layout built by Dave Frary and Bob Hayden, was covered in a popular series in *Railroad Model Craftsman* in the early 1970s. It is a fine example of "texture," that hard-to-define trait that blends all of the various models and details into a contiguous whole with no apparent discontinuities or anachronisms.

"Texture" as it applies to modeling is a term I first learned from Dave Frary, an old friend who wrote a best-selling book on model railroad scenery, *How to Build Realistic Model Railroad Scenery* (Kalmbach). He uses the term to describe the total scenic treatment – color, form, and texture. "I once called texture the key to realism," he recalls, "and after almost 35 years of scenery building, I find this still to be true." The key point is that the concept of texture on a model railroad goes far beyond surface roughness; it includes everything that helps to blend individual models into a contiguous whole (**3-1**).

Conveying the passage of time

Model railroads are among the best time-travel machines ever built. I can take a trip back to the Midwest of the mid-1950s simply by walking down into my basement. Things that are old now were new then; things that were old then are gone now. Today's UPS was yesterday's Railway Express Agency; today's Exxon was yesterday's Esso.

Capturing this sense of the passage of time guides our modeling endeavors. Very few things we see today were put there yesterday. Structures and other details ought to show signs of having waged battles with the weather, and of regular use and even abuse, for years if not decades – in other words, texture (**3-2**).

Signs of changing times include the scars of old stores on once-abutting buildings and foundations of demolished structures in the weeds (**3-3**). An abandoned or neglected building suggests key changes in a railroad's operating needs (**3-4**).

It is hard for some modelers to build or superdetail a model as neatly as they possibly can and then grunge it up to show the passage of time. The choice a modeler needs to make in this regard is whether the point of a model-building exercise was to build a model of a piece of railroad hardware or to model the day-to-day functions of a railroad.

An easy way to show a hint of yesterday is to spot some maintenance-of-way equipment on a little-used siding (**3-5**).

Enough – but not too much

A key aspect of texture is the need to slightly exaggerate what we'd actually see if an active object were reduced to our modeling scale. We innately know the aging wood siding on an old caboose or depot has surface texture, but it may be so small when reduced to HO or N scale that we can't see it at all. The resulting model thus looks a bit too neat and smooth unless we find a way to highlight the surface texture.

Dave Frary (www.mrscenery.com) has long advocated dry-brushing the edges of shingles and other raised surfaces to make them stand out (**3-6**). A

These photographs of a company house and barn illustrate that, beyond the basic structure, texture is often the most evident characteristic of a building. Not modeling the house's roof repairs and clothesline or the peeling paint on the barn would be to miss the essence of these structures.

3-2

little of this usually goes a long way, but not accenting the model's surface texture may fail to produce the desired visual effect.

It ain't on the level

A tip from Paul Dolkos: Create elevation differences, even on a "level" railroad (**3-7**). As you can see, he has elevated the main line and then varied the height of industrial spurs. Even raising the ballasted main line a scale foot above sidings improves realism.

Putting buildings and roads on hillsides (**3-8**) adds visual interest to a scene without taking much more depth.

Building in distance

Master modeler Frank Hodina painted some O scale Illinois Terminal equipment for a friend by matching the orange to restored equipment at the Illinois Railway Museum. "The models looked perfect outdoors," Frank said, "but the orange was too dark on his layout despite the fact he had very good lighting." The same concerns apply to scenery and structures.

NMRA Master Model Railroader Jim EuDaly, a retired optometrist, has a clinic on "scale color." He points out that military modelers long ago learned to adjust colors to look realistic at model-viewing distances by adding

Nothing that we regard as "permanent" happened a few seconds ago, so it should show some signs of change. The scar on the building wall (above) proves that what is now a parking lot once housed another building. The water tower footings in the weeds (left) similarly convey evidence of another time when steam power ruled the rails of the author's Allegheny Midland.

3-3

3-4

Depots fell into disuse as passenger trains were discontinued and agencies were consolidated, but some – including this 1974 example on a Louisville & Nashville branch – saw continued use and offer an example of how to add detail texture to a model.

white. "Daylight is 5,000 to 10,000 foot-candles," he said, "but my layout lighting is about 20 foot-candles. Less light means less reflected light from our models, so they should be painted lighter than their prototypes." As a starting point, Jim suggests adding 5 to 17 percent white for O scale models, 10 to 28 percent for HO, and 18 to 39 percent for N. Check the March 2005 *Mainline Modeler* or May/June 2005 *N Scale* magazine for more information.

Dave Frary commented that modelers should be prohibited by law from painting models with pure white or pure black paint. His point is that we need to build in a sense of atmosphere, as we need to think in terms of viewing our models from several hundred scale feet away, not an actual foot or two.

Bear in mind that applying a dull finish over a shiny surface often lightens the underlying color.

Achieving depth

When Dave Frary and Bob Hayden operated a commercial layout-building service, Dave typically built ⅛th scale models of proposed layouts (**3-9**), so he and clients could see in advance how the scenery would look. He scanned paper structure models from Dover cut-out books, reduced them, glued them to cardstock, and folded them to make scale-size buildings.

Dave uses as many different textures as he can. "To get depth, I apply texture over texture, gluing each layer in place before adding the next. The variety of textures includes commercial products, sifted dirt, bits of twigs, lichen, stones, and even real dead leaves."

Signs

State highway signs help identify the state where the railroad is located – two different states, in my railroad's case. They also hint at the era being depicted (**3-10**), and they simply add texture – richness – to a scene.

Highway signs have changed over the years (see, for example, www.ugcs. caltech.edu/~jlin/signs/). Yellow stop signs debuted in the 1920s and were replaced by red ones starting in 1954, although some lasted into the 1960s.

The U.S. Department of Transportation's standards for railroad signs and highway markers are available online at mutcd.fhwa.gov/pdfs/millennium/12.18.00/8.pdf.

Building signs offer another way to add texture to structures and scenes. Microscale has a number of useful sign decal sets, such as 87-795.

I often use a technique I picked up from George Sellios, of Fine Scale Miniatures kit fame, that makes it easy for a sign or logo cut from a magazine to appear to be painted right on a brick wall (**3-11**). (See, for example, www.billboardsofthepast.com, www.cocacolaclub.org/faq.shtml, and www.trevinocircle.com/adsigns.asp.) Vintage gas pump photos can be found by searching for "petroliana."

One caveat: Watch the typefaces. Helvetica, for example, is often used by modelers on old signs and even to letter or number steam-era rolling stock, as it's readily available in dry transfers and on computers – yet it debuted around 1964. A better choice might be an old standby such as Copperplate Gothic, which was the typeface of choice on railroad business cards and letterhead in the pre-diesel era, or even Arial, which is modern yet has the straight-legged R as well as a G that typify many older fonts. Compare those letters to the curved-tail R and the "goatee" on the G in Helvetica and many other modern fonts.

An easy way to ensure that your signs are correct is to search Web sites and eBay for photos of old signs (**3-12**).

Burma-Shave signs were once staples of road trips. They're considered collectibles today, and several Web sites such as www.lincolnhighwayassoc.org/iowa/shaver/1953.html, www.fifties-web.com/burma1.htm, and www.two-lane.com/burmashave.html show examples of those iconic signs and list all of the rhymes that appeared on them.

Need to verify the color and other details of license plates for your area and era? Check www.jbot.ca/plates. Need custom-lettered modern green highway signs? See www.kurumi.com/roads/signmaker/signmaker.html.

Easy "scenery" – a tender salvaged from an L&N Berkshire sits with other maintenance-of-way equipment near Pineville, Ky., in 1974. It inspired a similar scene on the author's Allegheny Midland. Total effort: a second knuckle coupler, decals, and a little weathering.

3-5

3-6

It's easy to add some texture to plastic "wood" surfaces by putting just a hint of paint on the tip of a stiff brush and drybrushing the color along the wood grain. The author is doing just that to a ready-to-use large-scale covered bridge he adapted for the Claremont & Concord project railroad he built for a *Model Railroader* series.

Three photos: Paul Dolkos

These scenes on Paul Dolkos' HO Boston & Maine show that varying the elevation of the main line compared to adjacent fields (above) or industrial sidings (right) adds considerable visual interest, hence texture, to a scene. The same thing applies to town scenes. Putting buildings on a hillside avoids the ping-pong table look (bottom).

3-7

Creative decal usage

Tom Johnson's HO railroad depicts a relatively modern era. To model soft-drink vending machines, he uses rectangular blocks of styrene covered with decals made from digital photos of actual machines.

John Nehrich photographed entire building sides, then made wall-size decals that captured all of the texture and detail of their prototypes. The decals were simply applied over blank wall surfaces (see the December 2001 *Model Railroader*).

Several companies now make plastic and paper products that have photographic images of tarpaper roofing, brick and rock walls, and painted wood walls, some with peel-and-stick backing. For example, check Clever Models (www.clevermodels.com/textures.html), Micro-Mark (www.micromark.com), and Paper Creek Model Works (www.papercreeek.com).

Posed people and vehicles

"Action figures" on a model railroad – people or animals apparently frozen in time as they are skiing down a hill, skating on a pond, swimming, or running – look a bit odd. The same applies to airplanes flying statically over the landscape or automobiles racing along an Interstate at zero mph. Instead, I model scenes that show people pausing for a moment to chat or rest (**3-13**) or vehicles waiting for a traffic light that never changes, and so on.

The visual interest created by town buildings on hillsides, as at Logan (top) and Hinton (above left) on the Chesapeake & Ohio, inspired this scene at Sunrise, Va. (above right), on the author's Allegheny Midland.

3-8

3-9

Dave Frary

Dave Frary built ⅙-scale mock-ups of custom model railroad layouts commissioned by clients so that he and they could visualize how everything would work as a whole before starting actual construction.

When this photo (left) of the Highway 49 crossing scene on Jack Burgess's Yosemite Valley RR appeared in MR, a California Hwy. Dept. engineer spotted the incorrect black background of the highway sign. Jack replaced the sign with a correct version the engineer supplied (above).

Both photos: Jack Burgess

3-10

Getting paper signs to nestle into the mortar joints between bricks like painted-on signs (left) is easy: Sand the back of the paper to make it tissue thin, trim it, brush on thinned white glue, and burnish it into the brick (right).

3-11

Speaking of vehicles, don't forget that new models used to be introduced in August or September, so having a shiny new 1957 Chevy (**3-14**) on a fall 1956 layout is appropriate.

Observing how people dressed in various periods is also important to help convey a specific time frame. Old magazines such as *The Saturday Evening Post, Look,* and *Life* are handy references and can be left on the coffee table in the crew lounge to add to the period flavor. Many Web sites such as www.sandiegohistory.org have vintage photos that show people, vehicles, signs, and other details.

Utility poles

In the long period prior to the advent of radio communications, the railroads were utterly dependent on telegraph and then telephone messages carried by copper wires. Railroads marked trackside poles at quarter-mile intervals so that engineers and conductors could time train speeds. Line poles were typically spaced 100 feet apart. (See sidebar on page 36).

Rob Enrico, who models the Pennsylvania in O scale, also takes great pains to model that railroad's signals and pole lines accurately (**3-15**). He feels strongly that such lineside details add considerable authenticity to his railroad.

Roads and highways

Pavement isn't quite the color we often imagine it to be. "Blacktop," for example, is only black for a short time after it comes out of the paving machine, and even then it's more of a dark charcoal gray (**3-16**). Soon thereafter, it lightens toward a medium- or even light-gray hue. For a brief history of blacktop paving, visit www.hotmix.org/history.php.

Fresh concrete typically has a gray coloration, although aggregate mixed with the cement bonding agent may tint the color toward red or several other hues. As it ages, its color changes as iron deposits within the aggregate oxidize, and as vehicles drip oil and grease, typically down the center line of each lane.

Both blacktop (macadam) and concrete roads develop cracks, which are sealed with shiny black tar. Like the oil drippings, these are easy to model and add considerable texture to a highway scene.

Paved lanes have gotten wider over the years. Whereas eight-foot-wide lanes (16-foot roads) were once common, especially for side streets, 10-foot lanes were often specified in the 1950s for public streets. Twelve-foot lanes are now the norm. The National Model Railroad Association (nmra.org) has published data sheets that provide specifications on a variety of such details.

It takes a little extra effort to make a convincing model of a gravel road. The original layer of crushed rock or gravel soon breaks down to a fine powder under ceaseless traffic, so the paths worn by the tires look lighter and smoother

3-12

Mike Sosalla

Jim Six photographed a lot of old signs, then improved them digitally and shared them with other modelers via the Internet. This Master Mix sign was printed on bright white paper and attached with cyanoacrylate adhesive to .010" stryene by North American Prototype Modelers club member Mike Sosalla.

3-13

Posing figures and non-rail machinery *between* spurts of activity is more plausible when they're viewed next to moving trains. Here a family takes a breather from garden work alongside the Norfolk & Western main line in southern West Virginia. They may remain in this "pose" indefinitely.

Certain automobiles, such as this restored '57 Chevy, are distinctive era markers. Put a model of this car in a scene and the modeled era can't be earlier than early fall in 1956 when the new model year began.

3-14

3-15

Rob Enrico

Rob Enrico sweats the details, including position-light signals and line poles, on his PRR layout. His O scale layout was featured in *Model Railroad Planning* 2006.

Pennsylvania RR line poles and wires

The lighter lines attached to the insulators on Pennsylvania RR line pole crossarms were used to operate the signals and the railroad's telegraph system and were also leased out to Western Union. I believe the heavier lower line is a coaxial cable used for telephone communication. The PRR usually had at least one of these lines, and sometimes many.

Berkshire Junction (www.bon.net/~kserre/berkshire-jon) makes a flexible "wire" product called EZ Line that comes in several colors,

including light green to simulate weathered copper wire. It is appropriate for the lighter wires attached to the crossarm insulators. As I explained in *Model Railroad Planning* 2007, I had O scale PRR-style line poles manufactured. I opted to string only the coaxial cable, which visually ties all of the poles together without the need to string 20 to 40 wires between each of the line poles.

I dye twine black by putting an entire roll in a jar of Rit dye, then attach it to the poles. The twine is heavy and assumes a natural

sag without any kinking. You could do the same in HO by using somewhat lighter twine. I drill a small hole in each pole just below the bottom crossarm braces, install a short length of .005" brass rod about ¼" long, and paint it the same color as the pole. The twine can then just be pressed on, as the brass pin will force itself through the strands of the twine. It can easily be removed and adjusted to get just the right amount of sag, then secured with a drop of CA.

– Neal Schorr

than the center and edges of the roadway. Just sifting fine sand or rock onto the road surface is therefore unconvincing, as the surface is too uniform.

I've had reasonable luck by sifting actual gravel found along country roads into thinned white glue applied to a tan-painted surface. After the glue has dried, I use medium sandpaper to remove the larger "rocks" from the more heavily traveled lanes of the roadway (**3-17**). The North American Pro-

totype Modelers group in Milwaukee uses natural materials for much of its scenery work (**3-18**).

Prior to the 1960s, road center line striping tended to be white (**3-16**), with broken lines marking the center and solid white lines (later yellow) denoting no-passing zones. I typically use white and yellow decal stripes available from Microscale. Dashed lines on two-way roads became yellow in 1971.

The stamped metal Armco roadside

barriers, made in HO by Rix Products, have been around since the early 1950s. They are typically given a silvery protective coating today, but vintage color photos show they were originally painted white with black posts.

Sounds and animation

We'll conclude our review of texture with a discussion of something you can't actually see – sound effects – as well as animation. I consider sound effects as much a part of scenery as anything we can actually see and touch. Among the very first additions to my in-progress NKP layout were bird and two-cylinder John Deere tractor sound modules (**3-19**) from Miller Models (www.millermodels.com). No matter what the weather outside, it's always sunny with chirping birds in the Indiana of my youth.

All of my locomotives are equipped with Digital Command Control sound decoders, but sound effects should not end with the locomotives. A model railroad with locomotive sounds but without background sounds is like listening to a singer performing a cappella. Only a very good singer can pull that off.

And just as viewing a play or opera performed on a barren stage could diminish the experience, you'd probably get less enjoyment operating a model railroad on bare plywood. Similarly, you shouldn't expect a model railroad to be at its best unless the sounds are as realistic and complete as the scenery, structures, and rolling stock.

For his HO Yosemite Valley, Jack Burgess plans to greet visitors with a sound and light show. "They will enter the darkened room and hear crickets and occasionally a bullfrog. A dog will bark in the distance. As the sun begins to rise, one will hear a rooster crow as building lights come on one by one. As trains begin to roll, the sky will brighten and normal daytime sounds will become the norm."

Animation is increasingly becoming an important aspect of convincing sound effects. It always has been part of onboard railroad sounds, but it now embraces not only that rotating carousel in the park but also operating rotary coal dumpers, preparation plants and coal docks that actually load coal, water spouts that raise and lower, even "big hooks" and container-loading gantries.

For more about the fascinating subject of animation, please refer to *A Beginner's Guide to Creative Effects for Your Model Railroad* by Paul M. Newitt (J-T Publishing, www.just-trains.com). You may also enjoy visiting www.fiftiesweb.com/fifties.htm.

Both photos: Jeff Wilson collection

Roads were striped with white paint through the steam era when yellow no-passing stripes began to be used, as these photos illustrate.

3-16

3-17

Gravel roads should show signs of wear where tires crush the aggregate, so the author glues down sifted dirt, then sands the traveled areas.

3-18

Mike Sosalla

George Thelen applies sifted real dirt to a scene on the North American Prototype Modelers HO layout in Milwaukee. The club uses "traffic bond" (crushed limestone) for gravel road surfaces.

Sounds beyond those built into locomotives are an increasingly important part of the texture of a model railroad. The author has equipped his farm-belt railroad with a variety of sounds ranging from chirping birds to a two-cylinder John Deere tractor sound chip from Miller Models.

3-19

4-1

CHAPTER FOUR

Rocks aren't hard

A layer of more-resistant rock caps the eroding mesa in the background in this evocative portrait of Western railroading on Paul Scoles' Sn3 railroad. He fashioned the rock faces by applying fast-setting dental plaster in rubber molds, as he explains in a DVD on making scenery.

As I stated in the introduction to this book, anything's easier to model when you have "plans" for the prototype. That's true when you're trying to replicate the look of a particular geographic area, and especially so if there are visible rock formations (4-1). A rock face that formed when sheets of lava flowed over or between sedimentary layers near the surface and rapidly cooled won't look anything like granite or sedimentary rocks. Similarly, a "young" mountain chain such as the Rockies (4-2) will only faintly resemble the ancient Appalachians, especially in the north where they've been "ground round" by bulldozing ice sheets. Learning how rocks formed makes it easier for us to model them.

4-2

A gentle upward-bowing in layers of sedimentary rock is clearly visible on this hand-carved cliff on Eric Brooman's Utah Belt. Eric varied the color among layers to represent different periods of deposition of the beds of sea creatures, sand, and mud that were later compressed into rock.

Types of rocks

I wrote an overview article about rocks and geology that appeared in the September 2005 *Model Railroader* titled "Rocks aren't hard." My point is that you can quickly learn enough about how rocks were formed, and later reformed and deformed, to model them realistically.

It's hard to model rocks realistically without a "blueprint." That blueprint is a general understanding of the basic types of rocks, how to tell them apart, and where they are likely to be encountered. Overlaying the route of your prototype-based or freelanced railroad on the accompanying geological province map (page 40) is a good place to start. No need to memorize it, but there will be an open-book test when you build the scenery on your model railroad.

There are three basic types of rocks: igneous (think "ignite") rocks, which were formed as magma cooled; sedi-

mentary rocks, which formed as earlier rocks were eroded by wind and water, built into layers, and compressed; and metamorphic rocks, which were originally igneous, sedimentary, or even previously metamorphosed rocks recast by intense heat and pressure.

For our purposes, we can make it a bit simpler than that: Either the rocks have a discernible geometric pattern – the distinct layers of sediments, no matter what their orientation (**4-4** through **4-16**), vertical columns of solidified magma (**4-17**), or "joints" that weather into distinctive blocks, to cite three examples – or they are massive monoliths of rock (**4-18**). The latter could be an outcrop of cooled magma or a single very thick bed of sandstone, limestone, or shale, but for all practical purposes, they are simply "big rocks." There are no clear layers piled one atop the other.

When we examine a rock outcropping, we should look for distinctive

geometric patterns or shapes. If we find them, we need to select rock castings or carve setting plaster into similar shapes using the actual rocks as a blueprint. Beds of sedimentary rocks separated by many yards or even miles may exhibit similar, repeating patterns of bed thickness and color (**4-12**), so care should be used when placing or carving the rocks. For crystalline rocks, a more free-form approach can be used, as there are not adjacent sedimentary bedding planes to align. This is where applying still-wet rock castings is practical.

The big picture

Viewing the continent in a very simplistic way – mountains or no mountains – is helpful. Where there are no mountains to speak of, as between the Appalachians and the Rockies, the surface rocks are usually sedimentary. You should expect to see occasional outcroppings of neatly layered rocks along streambeds and in cuts or the occa-

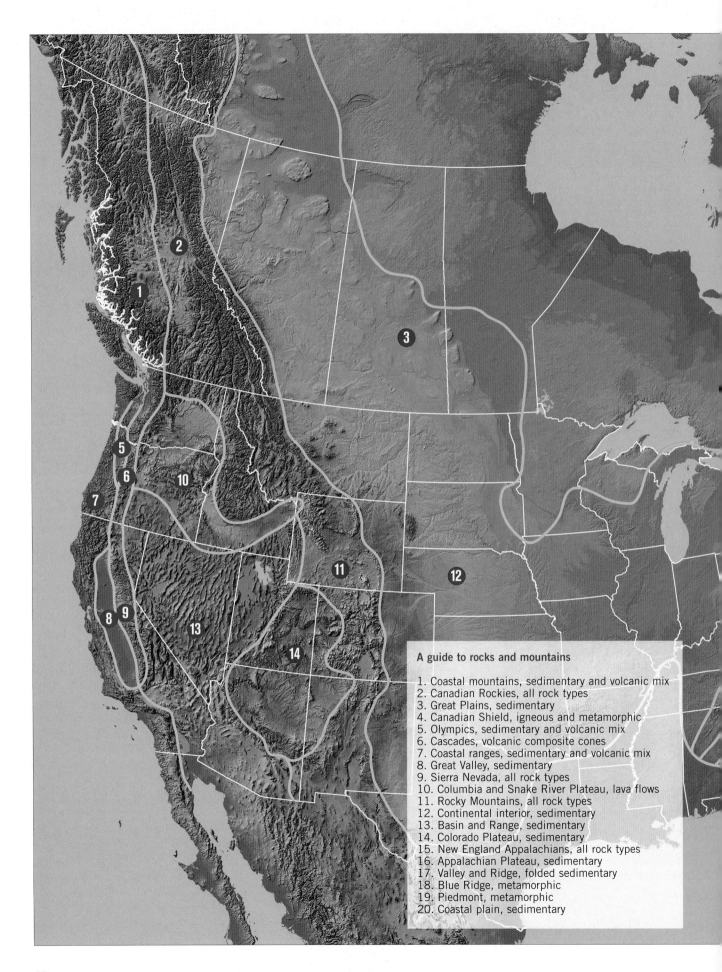

A guide to rocks and mountains

1. Coastal mountains, sedimentary and volcanic mix
2. Canadian Rockies, all rock types
3. Great Plains, sedimentary
4. Canadian Shield, igneous and metamorphic
5. Olympics, sedimentary and volcanic mix
6. Cascades, volcanic composite cones
7. Coastal ranges, sedimentary and volcanic mix
8. Great Valley, sedimentary
9. Sierra Nevada, all rock types
10. Columbia and Snake River Plateau, lava flows
11. Rocky Mountains, all rock types
12. Continental interior, sedimentary
13. Basin and Range, sedimentary
14. Colorado Plateau, sedimentary
15. New England Appalachians, all rock types
16. Appalachian Plateau, sedimentary
17. Valley and Ridge, folded sedimentary
18. Blue Ridge, metamorphic
19. Piedmont, metamorphic
20. Coastal plain, sedimentary

sional tunnel (**4-20**). They will usually be level or nearly so with some notable exceptions, such as the Cincinnati Arch, a gentle up-warping of sedimentary beds extending from Alabama through Ohio. Layered rock castings made from coal or shale masters or even ceiling tile will replicate this terrain quite nicely.

North of the Great Lakes in Canada, however, the ancient bedrock underlying the continent was exposed as thick ice sheets repeatedly scraped off overlying sediments. This extremely old and tortured metamorphic rock, called the Canadian Shield, typically displays no obvious layers, so crystalline rock castings are suitable. Stafford Swain described how he modeled this imposing topography in the January 1979 *Railroad Model Craftsman*.

In the West, the Rockies are part of the chain formed when the North American plate plowed into island arcs off the West Coast (**4-19**); see the following section on the Rocky enigma.

The East Coast

In the Appalachians, you may encounter all three types of rock, but most often, you'll spot layers of sedimentary rock or, especially in the Blue Ridge, the exposed continental metamorphic bedrock. No need to determine which is which or whether the sediments are shale, sandstone, coal, or limestone (**4-12**); all you really need to be concerned with is how they look in terms of layer thickness, color, and orientation (flat, tilted, folded, or monolithic).

The central and northern Appalachians were shoved up when the American and African continental plates collided hundreds of millions of years ago. This happened not once but at least twice, so today's Appalachians are really only an uplifted and partially eroded echo of the much higher originals. The first peaks were eroded down to a near plain, called a peneplain

Locating your railroad on this geological map of North America will provide clues to the type of terrain you should model. State geological maps and U.S. Geological Survey topo maps (see chapter 10) will provide finer detail.

("almost-a-plain"), which was later pushed up to form the summit of today's Appalachians.

The evidence is pervasive. Note, for example, that the verdant Appalachian ridges are all of roughly the same height (**4-21**). Rock types and layering, and the fossils they contain, match on either side of the still-opening Atlantic Ocean. Add to that the fact that the New River – ironically, one of the oldest rivers in the world – meanders like the old river it really is; it managed to cut through the layers of rising rock as fast as the new Appalachians were being elevated.

The Ozarks and other southern ranges (**4-9**) apparently endured a different scenario. One report postulates that a hunk of the U.S. Gulf Coast wound up stuck to Chile when North and South American plates were jockeying into today's temporary alignment.

These collisions compressed resting sedimentary rocks into sine-wave formations (**4-22**) or, at the very least, elevated them to form today's highlands and ridges. The pressure of the collision between continental plates was felt most acutely in the Northeast. Beds of soft coal were either eroded away or buried and folded into hard (anthracite) coal. Shale metamorphosed into slate, hence the Slate Belt in eastern Pennsylvania, and limestone into marble, hence the marble quarries in New Hampshire.

In the Appalachian Plateau, which lies to the west of the "hinge line" that angles northeast-southwest through central Pennsylvania and eastern West Virginia, you will find elevated but not folded sediments: limestone, sandstone, shale, and bituminous coal – and the railroads that made a living hauling soft coal to market.

East of the Appalachians is the Piedmont, a sloping plain of sediments from the eroding mountains that eventually dips into the Atlantic. There are extensive areas where "Triassic red beds" tint the soil a rusty hue, a characteristic usually associated with Georgia but also evident as far north as central New Jersey. Modeling this soil color complements other aspects of a scene set in the Southeast.

4-4

The distant future of this mesa is clear from the piles of rubble that flank it. The image is from May 1990 along the Denver & Rio Grande Western west of Helper, Utah.

4-5

An image from Amtrak's *California Zephyr* shows a mini-mesa about to disappear below a sea of sand of its own making.

That the remaining northern side of Castle Gate (right) alongside the Rio Grande (now Union Pacific) main line was once a sea floor is apparent from the horizontal beds of sedimentary rock. More-resistant upper beds have slowed but can't stop erosion even in an arid climate. The photo was also taken from the *Cal Zephyr* in September 2006.

4-6

Central North America

Loop around the south side of Chicago on the toll road and you'll find yourself driving on a narrow isthmus of rock between deep quarries. The limestone being extracted from these pits was formed when North America was turned "sideways" (today's north was then facing east) and straddled the equator, with the central part of the continent depressed and covered by a tropical ocean.

As seashells from countless dearly departed creatures rained down, layers of calcium carbonate built up and slowly turned to rock under the pressure of the water and subsequent layers of shells. Millions of years later, after the great ship North America had swung around and sailed north and its midsection warped upward above sea level, the limestone was within reach of highway and railroad builders.

Other limestone quarries, such as those in southern Indiana, were similarly formed. The stone was carved into beautiful monuments and buildings in our nation's capital – or, in terms more appropriate to our interests, became revenue-producing carloads for railroads.

The tropical climate was also good news for huge plants that grew into lush forests and then formed beds of carbon-rich material. It, too, was buried

4-7

4-8

The long horizontal lines of these Norfolk & Western Alco Century 630s complement the equally horizontal strata exposed behind them. Note varying thickness of beds in this November 1973 view south of Mullens, W.Va.

Sandstone and limestone beds are more resistant to erosion than soft shale and coal, which resulted in this cliff being undercut in the central Appalachians.

and compressed until many of the volatiles were driven out and it became coal, a rock that readily burns. Drive through southern Indiana and Illinois today and you will see a surprising number of coal-preparation and power-generating plants, as I documented in *The Model Railroader's Guide to Coal Railroading*. That's more good news for those of us seeking an interesting prototype to model in the nation's heartland.

In the northern U.S. and parts of Canada, rich iron ore deposits were discovered, chiefly in the fabled Missabe Range. These quickly eclipsed lower-grade deposits that had been mined as early as the Revolutionary War era in New Jersey. Even the genius of Thomas Edison couldn't compete with the Missabe, and his iron refinery near Sparta soon shut down.

Gravel pits (**4-23**) are common in the East, Midwest, and other areas of the continent that were repeatedly visited by ice sheets. The heavy ice ground rock outcroppings into smooth shapes and deposited the toils of its labors in elongated sand and gravel beds called eskers and lateral and terminal moraines as glaciers melted and receded. See, for example, www.geo.msu.edu/geo333/moraines.html and www.lgsb.uiowa.edu/service/geology.htm.

4-9

Most sedimentary rocks in the central U.S. remain horizontal even when uplifted, as in this Texas & Pacific scene from the author's collection. Note hints of lower layers of rock almost hidden in the eroded slope.

4-10

4-11

A large commercial rock casting (above) was painted a buff color and draped with Woodland Scenics foliage to create this scene along Coal Fork on the author's Allegheny Midland.

Note how the original surface of this rock cut on the Clinchfield (CSX today) near the Altapass Loops (left) continues from one side to the other.

This hand-carved cut on John Armstrong's O scale Canandaigua Southern (below) features sedimentary rock that has been uplifted but not folded or faulted. Rock that has broken off lines the stream.

The Rocky enigma

The Rockies are a series of north-south trending mountain ranges separated by wide, flat-floored basins, called parks in Colorado. Their cores are generally igneous or metamorphic rock (**4-17**, **4-18**, **4-25**). Modeling them is largely a matter of choosing large, structureless (no bedding planes) rock castings. The flanks of the ranges are characterized by sedimentary rocks raised and tilted as the Rockies were forming (**4-26**).

Any school kid can see that the continents on either side of the Mid-Atlantic Ridge fit together rather neatly and hence must have been joined at one time, so it's easy to imagine some turmoil when they bumped together and split apart. But west of the Rockies, there is no smoking gun, no opposing continent that played hit-and-run with our West Coast. Moreover, the Rockies are high and sharply defined, so they are relatively young – tens rather than hundreds of millions of years old. That's yesterday in geologic time. What had happened so recently?

Studies of the composition of the so-called "exotic terranes" that make up the West Coast seem to have unraveled

4-12

John Armstrong

the mystery. As the North American tectonic plate sailed ever westward, as it is doing at this very moment, it encountered the Pacific plate and island arcs. What had been oceanfront property in Elko, Nev., was the scene of a slow but monumental series of collisions.

Just west of today's Denver, once-flat prairie sediments were bent upward like the hood of a car during a head-on crash. Closer to the scene of the accident, the tectonic tangles formed additional mountainous heaps that became the Sierra Nevada, the Olympics, and the Cascades, among others.

West of the Front Range, you can find examples of the igneous basalts and granites that filled the void as the sediments were elevated, as well as the folded sedimentary rocks themselves. Each scene along, say, the Denver & Rio Grande Western (today's Union Pacific line through Moffat Tunnel and Grand Junction) should be tailored to match the rock types you see alongside the railroad.

Speaking of basalt (or rhyolite) and granite (or gabbro), it's worth gaining a rudimentary understanding of the difference. Both are cooled magma, but basalt (or, if more silica is present, rhyolite) forms closer to the surface as it cools more rapidly. The result is often a series of multi-faceted vertical columns. The Devil's Tower of *Close Encounters*

Bernard Kempinski

Bernie Kempinski modeled a sharply dipping sedimentary rock cliff along the Chesapeake & Ohio in N scale. Nearby rock outcroppings would show this same sequence of beds, although they could lie at different angles due to local folding.

A large syncline (rock strata with beds dipping toward each other) and fault were revealed by this cut in Maryland. Upper formations have been shoved to the left over other beds.

Rock castings made from molds of coal and shale create sedimentary beds that dip to the right on the Allegheny Midland.

The author started with a commercial foam rock casting, painted it black (above), then dry-brushed on a buff color to create this tunnel portal scene (right) at Coal Fork on his Allegheny Midland. The forest canopy is foam-covered poly fiber puffballs.

Vertical formations of rock can result when sedimentary beds are severely tipped or when magma cools rapidly. Joints in thick exposures of granite and similar rock may also give the appearance of "vertical" rocks, as do old necks of volcanoes, such as the Lizard Head formation along the Rio Grande Southern (right).

4-17

4-18

The famed Hanging Bridge in the Royal Gorge, now traversed by riders on the Royal Gorge Dinner Train out of Canon City, Colo., attests to the narrowness of this canyon in the relatively young Rockies. Little has changed since this Denver & Rio Grande Western publicity photo was taken.

4-19

The northern Rockies are also relatively young mountains but have been glaciated, creating U-shaped valleys and sharp-edged peaks. Here Canadian Pacific Extra 4037 East travels between the Spiral Tunnels on July 16, 1965.

Uplifted but not folded sedimentary rock characterizes much of the central U.S. from the Appalachian Plateau to the Rockies. Here a Detroit, Toledo & Ironton Geep emerges from a tunnel punched through beds of sedimentary rock to serve a brickyard at Fire Clay, Ohio, in May 1969.

4-20

4-21

The surface of the peneplain that developed after the original Appalachians had eroded away now forms the summit of today's Appalachians. This scene is along the C&O/CSX near Hawk's Nest, W.Va.

of the Third Kind fame is a good example of this columnar structure, as are the famous Palisades along the Hudson River. Yellowstone Park rocks are largely rhyolite.

When molten rock cools more slowly deep underground, it forms a very different type of rock, one familiar form of which we know as granite. Granite quarries can be found in various parts of North America; New Hampshire is one famous source of granite used for buildings and grave headstones. The Adirondacks are gabbro, which is chemically like basalt but has cooled more slowly deep underground like granite.

The West Coast
Looming between the fertile fields of California's Central Valley, discussed in chapter 8, and customers to the east were the Sierra Nevada ("snow-covered mountains"). The Central Pacific was the first to push east toward a connection with the Union Pacific in Utah, and in doing so, it became associated with names that have a magical ring to them, such as Emigrant's Gap, Norden, and Donner Pass. Central Pacific successor Southern Pacific turned Mallets around backward to put their cabs and engine crews out into fresh air as trains struggled through choking tunnels and snowsheds.

I recently visited the California State Railroad Museum in Sacramento to, among other things, pay homage to

This view of badly folded sedimentary rock along Route 23 in New Jersey (above) shows anticlines and synclines in close proximity. Gently dipping sedimentary beds east of Durango, Colo. (right), and at Sullivan's Curve in California (below) are more typical. Note thick beds of highly weathered sandstone in the photo of the Union Pacific eastbound freight.

Jim Boyd 4-22

Visible effects of glaciation range from well-rounded peaks where ice sheets bulldozed over them (the Delaware Water Gap between Pennsylvania and New Jersey, above) to sand and gravel beds along the sides and front of decaying ice sheets. Higher mountains like the northern Rockies and Alps are sculpted into sharp Matterhorn-type peaks by ice accumulation (right).

4-23

4-24

Yosemite Valley (left) is a classic glacier-carved U-shaped valley. Here's Half Dome as seen from Glacier Point in May 1990.

By contrast, the Arkansas River valley near Tennessee Pass retains the V-shape of a young stream that hasn't been scoured out by glaciers.

4-25

the sole remaining Cab-Forward, class AC-12 4294, and to visit the original Central Pacific shop buildings that are slated to become part of the museum. The battle between man, machine, and mountain has been and remains a heroic struggle, and those who honor it by modeling the Central Pacific, Southern Pacific, or successor Union Pacific through that region should complement the machinery with equally well-detailed scenery (**4-27**).

To the north, the railroads that form today's Burlington Northern Santa Fe, especially the Northern Pacific, Great Northern, and Spokane, Portland & Seattle, had their own burdens to bear. The UP and former SP&S parallel either side of the Columbia River Gorge, which was formed when waters emptying from a huge glacial lake cut into an extensive lava flow. Similar obstacles faced the builders of the Canadian transcontinental railroads (**4-19**).

Columnar basalt like that along the Columbia River's flanks is challenging to model. The columns could be carved into setting plaster like any other rock, or blue or pink foam could be carved and coated with a thick wash of plaster. Or try carving square balsa strips into crude hexagonal shapes, making a rubber mold, and casting multiple copies from pre-colored plaster. Break them into sections a few inches long and then restack them randomly to create the basalt columns.

Read the instructions

Modeling mountains or rock outcroppings accurately isn't hard if you follow the tongue-in-cheek advice of my friends who are design engineers: MILTFP41. That is, make it like the … plan for once.

If you know what rocks from various regions of the continent should look like, you'll know what to watch for along your railroad's right-of-way, which gives you strong clues as to how to model them.

Accurately modeling rocks and landforms is no different than modeling any other structure. Just follow the prototype plans and photos.

4-26

Bruce Meyer

Electro-Motive's Bruce Meyer photographed the sharply dipping Front Range sediments near Tunnel 18 from a new Rio Grande GP30 in May 1963. The once-horizontal rocks were formed under an ancient sea, then elevated to become part of a mountain range.

4-27

Jim Boyd

Mountains west of the Rockies vary greatly in appearance, from towering mounds of granite to still-active volcanoes, so careful observation of the rocks and vegetation of each region is required to model it effectively. Southern Pacific 4449 leads a fan trip at Lake Shasta, Calif., in May 1984.

5-1

CHAPTER FIVE

Water's ways

A workhorse 2-6-6-2 Mallet trundles back from the mines and across Coal Fork on the author's Allegheny Midland. The creek and rapids were formed with wet plaster and rock castings, then painted with acrylics and coated with gloss medium.

Water and rock are mortal enemies. Water, soft as it seems, always wins eventually. Its role in shaping landforms means that water and evidence of its work cannot be ignored as we scenic our model railroads.

Water can assume all sorts of colors and surface textures depending on whether it lies peacefully in a pond, rushes madly from up here to down there, meanders back and forth along an ancient streambed, crashes into jetties and sea walls, or reacts to the passing of a vessel. Sometimes it's almost transparent; at other times, it's absolutely opaque. It follows that modeling water requires different techniques and materials to achieve the desired final result.

5-2

The slightly rippled surface of the river still reflects a Western Maryland bridge at Elkins, W.Va., making this scene a good candidate for modeling with casting resin.

One size doesn't fit all

Since there are so many types of waterways and basins, there's no one best way to model water. What works for an Appalachian stream (**5-1** through **5-3**) may look wrong for a scene in New England (**5-20**) or Colorado (**5-22**).

Scale effect

One additional caveat about modeling water before we look at some specific examples and techniques: Water has a "scale effect." That is, if you fill a large dish with water and look carefully at it, you can sense about how far away your eye is from the water – a few inches, perhaps a foot or two. That's one of several reasons why using real water on a model railroad isn't a good idea: It doesn't appear to be far away. (It also elevates the room's humidity, can leak, is very heavy, and poses other practical problems.)

The same problem can occur with epoxy resins; they can appear too smooth, and your eye can sense that it is seeing a surface that is mere inches, not hundreds of scale feet, away. This may work in a small pond, but it can be distracting in a large harbor scene.

Lance Mindheim suggests applying a coat of gloss medium over the resin so that the surface isn't unrealistically smooth. This is also a good way to renew the gloss of an older surface.

The moral of the story is not to

5-3

Before and after photos show how the pea-soup-green stream and low falls over sedimentary ledges were transformed into a rushing rapids with streaks of titanium white acrylic and gloss medium. (See the cover for another view.)

5-4

Note the contrast between the still water above the rapids and the turbulent water in and below them in this scene on Dan Zugelter's HO Chesapeake & Ohio layout.

5-5

A similar scene along the C&O's Greenbrier Branch ranch to Durbin, W.Va., photographed on a fan trip in May 1971, shows areas of glassy and rippled water.

5-6

Bob Hayden

Bob Hayden used resin to form a reflective stream on his HOn2½ Carrabasset & Dead River RR, a free-lance railroad based on the real Maine two-foot-gauge railroads. He strongly recommends testing each batch of resin to be sure it will set properly.

avoid using epoxy resins for water but rather to be aware of the desired effect you want to achieve. Then choose the materials and techniques you will need to achieve the desired outcome. When in doubt, build some small scenery test sections and experiment until you are sure you know what you're doing, and that the materials you have on hand are up to the job.

Quiet water

When considering modeling water, the first thing that pops into many modelers' minds is epoxy casting resin. That's not surprising, since it really does look wet after it dries. When used to model ponds and relatively smooth or transparent streams (**5-4** through **5-8**), it's hard to beat.

An ever-growing variety of products nicely simulates quiet water. Casting resin can be a bit tricky to use: Epoxy is self-leveling, so the streambed must be absolutely flat, and epoxy will find the tiniest of holes and leak onto the floor. If poured in one thick coat, it may crack. If surface texture is desired, epoxy must be "worked" as it sets. It can creep up onto the shore or walls, so brushing on "stop" lines is often helpful. It's not a good idea to use epoxy that has been stored for a while, as it may never fully harden; always mix a test batch to be sure it will set up properly. And pre-pour preparation is everything, as you can't go back and start over. But, done correctly, the effects can be dazzling.

Still water can also be modeled with other liquids such as Realistic Water from Woodland Scenics, Lakes 'N' Rivers from Model Builder's Supply (www.the-n-arch.com), or even sheets of glass or plastic.

Rivers and creeks

I grew up in the Midwest and spent a lot of time hiking along the banks of the Wabash River in Indiana. As it heads south toward the Ohio and then the Mississippi, it carries a lot of sediment, especially after a heavy rain or during the spring thaw. It usually has a pea-soup green appearance, sometimes tinged with brown, and it is absolutely opaque – the underwater portion of a tree limb that

5-7

5-8

Glassy settling ponds, like this one at the Summerlee (Lochgelly) prep plant on the former Virginian Ry. near Oak Hill, W.Va., in November 1973, are good candidates for poured-resin water.

Small reed-lined and weed-choked ponds are also ideal for resin water. They're small enough to lend themselves to superdetailing with various types of ground cover.

5-9

5-10

Most Appalachian rivers and streams are relatively opaque and usually pea-soup green. Note the gently dipping sedimentary rock formation, which has caused the stream to tumble over a submerged ledge. This scene is along the Clinchfield.

A dam at the Baird farm along the Lehigh & Hudson River south of Warwick, N.Y., shown in November 1971, creates a glassy pool upstream and foamy streaks below the dam.

5-11

5-12

The trunks of trees along Appalachian rivers and cuts are seldom visible, which saves lots of modeling time – puffball trees are excellent for modeling these forests.

A similar trunk-less tree effect is evident along the Mississippi River in this photo of a Burlington Northern freight north of Savanna, Ill., in June 1993. Sky reflection makes the water appear blue.

5-13

The blueish cast to the river was modeled by referring to an inspirational prototype photo by Don Valentine. This scene is on the author's large-scale Claremont & Concord project railroad.

5-14

The water on the C&C was modeled using a thin layer of plaster over the foamboard base. It was painted with acrylics, then given several coats of gloss medium.

has fallen partially into the river might as well be cut off at the surface.

The same is true for many of the rivers I saw during myriad photographic trips to the central Appalachians (5-9, 5-10). I also noticed that leaves of trees lining the riverbanks tended to dip into the water. Their trunk and branch structures were all but invisible (5-11), a trait not reserved

to Appalachian forests and streams (5-12).

Such streams are easy to model by pouring about ¼" of soupy plaster into a prepared streambed and, when thoroughly dry, sealing it with latex paint. Color the surface with acrylic paints (5-13, 5-14), then add two or three coats of gloss medium to achieve a wet look. If the surface gets dusty or dulls over time,

simply clean it off with a wet sponge and add another coat of gloss medium.

To get a rougher surface, Michael Pennie applied gloss Mod-Podge with a small brush, working slowly to avoid air bubbles (5-15). "The waves just happen by making them as you go," Michael said. "If you get tired, you can come back to finish the scene later, as it's easy to blend with what's already done." He suggests an overcoat of high-gloss urethane or gloss medium if more shininess is desired.

In New England and some other regions, however, streams tend to be much more transparent (5-16). Here, epoxy resin or another dry-clear product clearly has the edge. In chapter 1, Doug Tagsold described how he used Enviro-Tex resin and clear and white silicone caulk to create waves and rapids on a section of Colorado's Animas River for his On3 layout.

The river barge's foamy bow wake in the muddy Mississippi on Allen Keller's layout adds a sense of movement to the scene (5-17). Even a barge or boat securely moored alongside the shore with rope lines or steel cables would still create a substantial wake owing to the fast river current.

Rapids and waterfalls
Soupy plaster can be made to flow around rock outcroppings and ledges

Michael Pennie achieved a ripple effect by coating an acrylic-painted base with gloss Mod-Podge. "Waves just happen by making them as you go," Michael reports. You can blend new work with old.

(5-3). I paint the dry plaster and rock castings with green latex, later go back and color the rock outcropping with tan or gray acrylic, then dry-brush titanium white acrylic to simulate the foam that forms downstream of anything interfering with the water's flow at or near its surface (see cover). Two or three coats of gloss medium add the wet look, and the water can be "renewed" at a future date with another gloss coat or two.

Such effects can be done with resins, but remember that they flow freely and will drain from the higher parts of a stream to the lower areas. Each level between the rapids or above a waterfall therefore needs what amounts to a dam – the rock outcropping – to pond the water (resin) upstream until it starts to set. As it sets, it can be worked over and around the ledges that create the rapids.

C.J. Riley has found that Woodland Scenics Water Effects produces better results than silicone caulk when modeling rapids (5-18). "Its applicator makes it easy to place the material around rocks and debris," he reports, "and it's easier to shape. I like it better than their water material for this purpose."

C.J. also prefers Enviro-Tex with gloss gel brushed over it for ripples and rough water. "At the bases of waterfalls and series rapids, I have had success

Resin was used effectively on the New England, Berkshire & Western HO layout to model the nearly transparent streams in Vermont.

with any crystal such as diamond dust or fine colorless glitter to provide the glint of real water. I secure it around boulders with small amounts of diluted white glue or even CA."

Waterfalls (5-19) are a bit more challenging. I modeled a low waterfall over a concrete dam using plaster, just as with rapids, but higher waterfalls need a sense of wispiness and transpar-

ency. There are several commercial products intended to ease this chore that look good when seen in a still photograph. Many modelers squeeze out beads of clear silicone sealant and work it smooth over the crest and vertically down the waterfall with an old knife.

My concern is how to convey a sense of motion to waterfalls. Playing a

5-17

Lou Sassi

Allen Keller modeled a small portion of the Mighty Mississippi and its barge traffic by spreading plaster unevenly on the base. He wrapped boats and barges with plastic food wrap, then built up a wake around each model. Allen used raw sienna and burnt umber acrylics to get the muddy color and created highlights by adding more raw sienna. Coats of gloss medium went on last.

recording of the roar of falling water will certainly help. I've seen such effects in artwork sold in mall kiosks, so adapting that technology to a model railroad might prove rewarding. In *A Beginner's Guide to Creative Effects for Your Model Railroad*, author Paul Newitt discusses using polarized light to create a moving-water effect.

Tidal water

The gravitational tug exerted by the moon on great bodies of water causes the daily tide cycle. In areas such as the Bay of Fundy in the Canadian Maritime Provinces, narrow passages between landmasses constrict the tidal inflow to the point where you can actually witness a tidal bore – a wall of water surging upstream as the tide advances.

Of more immediate interest to modelers are the visible effects of tide levels on shore installations such as the seawalls around coastal harbors. Dark stains and even barnacles show the maximum normal height of the tide (**5-20**).

The oceans can be glassy smooth, but most often waves are formed by wind blowing across their vast surfaces. As the waves approach shore, the ever-shallower depth drags on the lower layers of water, causing the waves to break as they crash on shore.

I've never seen a model of a breaking wave, but I have doubts about its

5-18

C.J. Riley

C.J. Riley used a sheet of clear plastic painted blue and shaded to sand on the underside. Gloss medium was applied on top. The rapids are clear styrene with clear silicone applied in multiple thin bands. White glitter was used to add highlights at the base of the falls. In other locations, he poured casting resin in several thin layers.

5-19

These falls along the Western Maryland's line north of Elkins, W.Va., shown in May 1973, are a good candidate for modeling using clear plastic sheet and beads of clear caulk. Sound effects would add to the visual drama.

effectiveness anyway; it would look good in still photos but strike the viewer of a model railroad as something frozen in time. Model railroads are about motion, after all; they are not usually dioramas or still-life displays. That's why figures (people) posed in the act of running, vehicles on busy highways, or airplanes suspended above the landscape may seem unrealistic compared to the moving trains.

But the waves moving into a harbor or evident in a river can be modeled effectively. Neal Schofield poured Woodland Scenics Realistic Water onto a sheet of Plexiglas, which helped to support a trestle (**5-21**) while minimizing the amount of water material required. The background scenery, by the way, is a photo taken in the Connecticut River valley in Vermont that was printed on a 24" x 36" sheet of photo-quality paper. The sky was cut out before it was mounted on the backdrop.

5-20

Dave Frary

Dave Frary gave viewers strong hints that the tide is still out – note the stains on the pilings and tall footings under the building. The scene says "coastal New England" without a word being spoken.

5-21

Lou Sassi

Neil Schofield achieved an excellent blend of 3-D scenery and a homemade photo backdrop on his HO Connecticut River Line. Neil used several coats of Woodland Scenics Realistic Water poured on a piece of Plexiglas. He created the ripples by dabbing the partially cured "water."

5-22

Young streams have steeper gradients and are therefore more turbulent. The author photographed this scene from the Royal Gorge Dinner Train in September 2006. Model rafts and rafters are available commercially.

Old vs. young streams

Young streams are those that result from a recent elevation of a landmass, "recent" being defined in terms of a few million to perhaps a few tens of millions of years old. The higher the elevation above sea level, the more potential energy the stream has to cut through any obstacle between it and tidewater. The resulting streambeds are narrow and steep, with turbulent water "rapidly" sawing away at the underlying rock (5-22).

The Rockies are a relatively young mountain range, so the rivers and creeks that drain them display the signature V-shaped valleys and relatively straight courses. Exceptions are found in the northern and Canadian Rockies, where recent (about 10,000 years ago) ice sheets scoured out the valleys into U shapes, complete with hanging streams that once merged effortlessly into a river that is now hundreds or thousands of feet below.

As streams age, they start to meander to create S-shaped streambeds (5-23). Lance Mindheim has done an exceptional job modeling this effect on his N scale Monon Railroad (1-13), as he described in "Modeling a creek" in the July 2003 *Model Railroader*. Details include the typical sandbars (5-24) and bank undercuts.

5-23

Over time, streams, such as this creek along the Cumbres & Toltec narrow-gauge tourist line, tend to meander from side to side.

The Mississippi is replete with now-isolated crescent-shaped lakes and ponds where the river once threw out its hip, then cut through intervening banks as it kept rolling along toward the Gulf of Mexico. Look carefully at topographic maps that show almost any older river and you'll find evidence of previous paths.

Old riverbeds may ease the task of railroad planners. The Chesapeake & Ohio's (now CSX's) route from Huntington to Charleston, W.Va., for example, follows the now dry bed of the former Teays River. Most of its path across Ohio was buried during the last Ice Age, with the Ohio River assuming its drainage duties.

Ditches

The civil engineers charged with keeping the railroad in running condition in good weather and bad live by a three-word credo: Drainage, drainage, drainage! If water is not channeled along and away from the roadbed, the substructure will fail. Ditches and culverts are therefore critical to a railroad's physical well-being and should be modeled.

The drawing in **5-25** shows a typical roadbed cross-section, including the ditches on either side of the subroadbed. The ditches are often dry and may be so weed-grown as to be invisible. But near tunnels, which tend to drip with ground water, the ditches usually are partially filled with water. This type of water is most easily modeled with resin, but again be aware that it has a tendency to creep up the sides of the ditch. Covering the "creep" with a second layer of ground cover usually cures the problem.

Work from photos

The main theme of this book is to work from prototype rather than model photos when building models. A model of a model is at best a second- or third-generation copy, and something is almost always lost in the translation. That applies to water as much as to structures or rock formations.

Jeff Wilson collection

Whether in the West (top) or East (above), streams deposit sand on the inside of their meanders, as that's where the flow slows down. Modeling sand bars adds considerable realism to a stream.

5-24

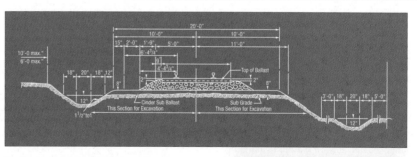

Drainage is a top priority to ensure a railroad's continued viability, and each company published roadbed cross-section drawings to document precisely what was required (above). This July 1974 photo of the Louisville & Nashville switchback and tunnel at Hagans, Va. (right), shows evidence of recent ditch clearing.

5-25

6-1

CHAPTER SIX

The forest or the trees

The ridge that rimmed the west side of South Fork, W.Va., on the author's Allegheny Midland was covered with puffball trees that became smaller toward the top to add an illusion of greater height. Scale-height trees, like the Woodland Scenics ready-made tree in the foreground, were kept well away from the ridge.

The old saw about not being able to see the forest for the trees certainly applies to modeling scenery. We're seldom modeling both at once: Either we're modeling a forest, or we're actually modeling individual trees. I mainly modeled forests on the mountainous Allegheny Midland, but I'm mainly modeling individual trees on my new flatlands railroad. The difference in these two approaches from both the standpoint of techniques and of results is dramatic.

This scene along the Clinchfield (CSX today), photographed in September 1975, shows how tree foliage completely obscures the branch and trunk structure except for a few small individual trees.

Trees as yardsticks

If we are modeling an individual tree or two, they can stand as tall, or almost as tall, as their living benchmarks. A single tree can be an impressive addition to a home's front or back yard, for example, perhaps sporting the tire-on-a-rope swing or tree house we fondly remember from our youth. A row of trees can also help disguise the lengthy horizontal joint where the landscape abruptly ends and the vertical sky backdrop begins.

Such trees become a liability, however, when we are trying to model a verdant ridge. That ridge may actually tower hundreds or even thousands of feet high, yet our model version rises only to perhaps 100 or 200 feet – a rather puny obstacle to man and machine.

Job One is therefore to extend the apparent height of the ridge with a bit of visual trickery, namely increasingly small, trunkless trees (**6-1, 6-2**). Even a single tree placed in close proximity to the ridge will blab to one and all that the king is wearing no clothes – that our mountain is little more than a mere molehill.

When I built the town of Coal Fork

as an extension to the Allegheny Midland, I used some scale-height trees in the foreground next to a row of company houses (**6-3**). There they added needed texture (see chapter 3) to the bare-bones homes. But they were used only between the fascia and the railroad branch, and well below the elevation of the railroad tracks, never on the backdrop side of the railroad.

Between the railroad and the backdrop was a steeply sloped ridge made of slabs of 2" blue foam board (**6-4**). Poly fiber "puffball" trees on toothpicks and a handful of foam rock castings were sufficient "detail" to convey the feeling of a towering ridge on one side of the valley carved out by Coal Fork.

At the base of the ridge near the tracks, the puffball trees were about 3" in diameter. As I added trees higher up on the ridge, I decreased the diameter to perhaps half that size.

Making puffball trees

The trick when making puffball trees from poly fiber material is to use very little of it per tree. Simply pull minute tufts of it off the blanket of raw material until a gossamer-like wad forms in the palm of your hand (**6-5**).

Gently roll it into a ball, then stretch it into a pinecone shape and toss it into a grocery bag. Fill a bag or two while watching TV.

Work outdoors or don a good spray-painting (respirator) mask (a dust mask is not sufficient) as you hold each puffball with a pair of long tweezers and spray it with cheap hairspray, spray glue, or flat-black paint. Dip the puffball into a coffee can filled with coarse turf, and shake the can until the armature is completely covered with foam. Toss the coated tree into a box. Any foam that comes off can be reused later.

I almost always use only dark-green coarse turf for forests, since you view them from a distance, and atmospheric haze tends to mute out any color variations between trees. Moreover, using a variety of foam colors can create a salt-and-pepper effect, and the lighter shades seem to stand out too much in photos. But viewed close up, some highlights may work to advantage, as is evident in Mike Garber's photo (**6-6**). Mike also prefers to put the poly fiber in place on the mountainside before applying the foam coating.

I planted the trees using double-ended wood toothpicks stuck first into

6-3

Scale-height (if a bit small) trees in the foreground, made using sagebrush trunk armatures, complement the company houses at Coal Fork on the author's layout. However, these trees would greatly diminish the apparent height of the background ridge if placed between the tracks and the mountain.

6-4

▲Slabs of 2" blue and pink foam were laid at an angle to the backdrop and covered with puffball trees on the author's Allegheny Midland. The foam could be lifted out to gain access to hidden staging tracks along the wall.

6-5

Poly fiber "puffball" trees start with a gossamer-thin ball of fill (left). They are then sprayed with hairspray, spray glue, or black paint before being coated with dark-green coarse turf.

the puffball and then into the scenery. You should be able to cover a two-square-foot area with one night's worth of trees, coat a new batch, and make an equal number of new puffballs in one evening. It took me about a week of evenings to cover the ridge behind each mountain scene.

Tom Johnson also uses poly fiber material to model the tangle of weeds along the right-of-way (6-7).

Built-in shadows

When I was building the Coal Fork Extension to the Allegheny Midland, I tried a trick that the late Jim Teese taught me. Instead of using spray glue to hold the foam foliage in place, Jim suggested spraying the puffball with cheap flat-black spray paint. "The paint costs less than the spray glue," Jim said, "and the black gives you a sense of shadow."

Jim also suggested using inexpensive white fiber material sold at fabric stores as pillow filling. I found painting the white fill material black worked fine for background trees, but in the foreground the foliage was sometimes rubbed off as crew members went about their chores, revealing the white material.

Black poly fiber is also available, as it's used for Halloween masks. One source, Putnam Co., can be found at www.softshapesdirect.com; click on Special Edition Fiberfill at top-left. Also watch for black fill used for Halloween costumes.

Individual trees

We've been discussing forests, so now let's look at individual trees. What color is a tree? It's a trick question, one that contains an important lesson. Let's look at the deciduous trees in **6-8** to make a point. You may be surprised at the answer.

Photo **6-9** shows three similar trees, one covered entirely with green foam foliage, the second covered with a dark green undercoat of foam that has been highlighted with lighter green to achieve the sun-shadow effects. The third tree has yellow foam dusted on to suggest strong sunlight from one side, or perhaps a hint of early fall. The trunks are gray, not brown.

One of the best series on modeling trees I have ever read was written by Don Ledger; it began in the November 1999 *Mainline Modeler* and continued into 2001. Among several other articles worth reading are "Modeling trees" by Bob Hundman (February 1999 MM), "The art of making trees" by Ken Patterson (November 2001 MM), and "Modeling bare trees" by Bill Henderson (January 1997 *Railroad Model Craftsman*).

Even when we're thinking in terms of individual trees, it's important to realize that trees tend to grow in clumps rather than singly. This is important to consider when dealing with the long, highly linear scenes typical of a model railroad, especially one built along a shelf. An odd tree here and there will look, well, odd.

Trunk and branch armatures

Many years ago, Woodland Scenics introduced bendable cast-metal armatures for the main trunk and several branches. Since that time, flexible plastic armatures from Woodland Scenics and other manufacturers have added to our options.

Tree armatures made from natural material such as sagebrush and even weed-like materials are available from Scenic Express (www.scenicexpress.com) and other suppliers. That company's SuperTrees and SuperSage tree armatures have gained wide acceptance. I frequently refer to their catalog when deciding how to scenic a given scene.

Both photos: Mike Garber

Mike Garber and Howard Heltman used black poly fiber (sold as Halloween Freaky Fluff by Union Wadding of Pawtucket, R.I.) to make forests for Mike's HO Virginia Southern Ry. They coat the hillside with Elmer's glue, spread out a wispy layer of fill, and press it into the glue. When dry, they tease the fill to gain height and texture, coat it with spray glue, and sprinkle on a 30- to 40-percent coating of dark green coarse foam, then add light green to about 90-percent coverage (so some black fill still shows). **6-6**

6-7

Jim Six

The tangle of bushes along the right-of-way was nicely modeled by Tom Johnson, who used poly fiber in two ways. In some areas, he stretched it out, sprayed it with hair spray, and applied ground foam (like making a tree). In other areas, he glued the poly fiber directly to the scenery base, applied ground foam, then sprayed it again. The weeds are Prairie Tufts from MiniNatur. Jim Six detailed the HO New York Central Geeps.

6-8

These late-fall deciduous trees show a common characteristic: gray trunks and branches. The cloudy-bright day softens shadows, much like fluorescent lighting. To simulate a sunny day, we'd have to build in shadows with dark under-foliage.

Coating trunk/limb armatures with foliage is usually done by spreading poly fiber netting, with foam already embedded in it, over the limbs. Another interesting technique, shown to me by John Bussard, is to cut untreated hemp rope into piles about ½", ¾", and 1" long. The trunk armature of choice is painted, then coated with a quality spray glue. A handful of the longer fibers are pushed into the armature; those that don't stick are reused on subsequent trees. Then more spray glue and progressively shorter fibers are applied. Finally, the entire tree is given a light coat of gray paint and dusted with the leaf material of your choice (**6-10**).

The recent trend toward ready-to-use models has spread to trees. Instead of looking for generic trees or even "deciduous trees," you can now get somewhat picky and look for a tower-ing oak, a bushy maple, or even a birch. I suspect a tree expert could pick some nits here, but for most of us, the variety alone is sufficient to add some "texture" to our scenery.

If you want to see some spectacular models of trees and New England landscapes as they evolved, visit the Harvard Forest Dioramas at the Fisher Museum in Petersham, Mass. You can learn a great deal about how trees look and where they naturally occur in *Trees of North America* by Alan Mitchell (Thunder Bay Press).

Commercial etched-leaf trees of several varieties are available from Samtrees (www.samtrees.com); the company also offers other types of trees including plastic.

Foliage options

It's impossible to keep up with the steady stream of tree-making products coming onto the market. Getting copies of the current Walthers and Scenic Express catalogs is your best bet.

I can recall the days when we used sawdust that had been dyed green for both ground cover and foliage. Then came ground foam, which was introduced by Architectural Scale Models and later made more popular among model railroaders by Woodland Scenics. Foam is still as popular as ever for foliage and ground cover, especially when glued to poly fiber armatures, but there are some options.

Some "leaf" material works well, especially in the larger scales, so it pays to experiment. When making large-scale trees for an MR project railroad, for example, I used Selkirk Scenery Co. (selkirks@telusplanet.net) leaf material.

Commercial products have recently expanded to include etched-brass sumac bushes (**6-12**) from Alkem Scale

6-9

Three tree armatures have been covered with green poly fiber and then, from left, very dark green, lighter green, and finally yellow foam to simulate shadows and sunlit highlights.

Models. Alkem produces a variety of etched-brass detail parts ranging from cornstalks to switch stand targets.

Modeling autumn trees is discussed in chapter 7.

The warmer climes

Popular as mountain railroading is with model railroaders, some have chosen to depict warmer and/or more arid climates. David Barrow's original Cat Mountain & Santa Fe depicted the sun-seared west Texas environment (**6-12**). In the December 1983 MR, Mark Junge offered a detailed overview on modeling the great American deserts, including ways to make distinctive native plants such as saguaro and barrel cactus, ocotillo, Joshua trees, palm trees, and various other trees and shrubs.

Pelle Søeborg has received wide acclaim for his superb depictions of scenery in the Mojave Desert. He described how he models desert scenery in his book *Mountain to Desert* (Kalmbach).

There are several sources for palm trees, including Samtrees, which makes etched-metal-leaf trees of many types.

Ground cover

When we look at a prototype photo of a railroad scene, we usually focus

6-10

John Bussard made this tree in just a few minutes from a commercial armature and short lengths of untreated hemp rope. He cut the hemp into short lengths, sprayed the armature, and dabbed small piles of hemp into the branches until a "crown" was built up. He then sprayed the tree gray and added ground foam for leaves.

6-11

Bernard Kempinski

Beautifully detailed sumac bushes are available as brass etchings from Alkem Scale Models in HO. Alkem also produces cornstalks, switchstand targets, and other detail parts using the same process.

6-12

On an earlier iteration of David Barrow's Cat Mountain & Santa Fe HO layout, Dave's wife, Lu Ann, painted the sparse vegetation typical of the west Texas landscape, which greatly expanded the apparent depth of this narrow shelf.

6-13

California's "gold" was mostly dried grass rather than ore, as here along the Union Pacific (former Western Pacific) in August 1992. Clearly, a wide range of tans and browns will be needed to replicate the hillsides.

on the locomotive or train. But even when the railroad's property is neatly groomed, as it often was during the steam era, natural chaos reigned on either side of the right-of-way. Weeds and shrubs of all description were in evidence.

Trying to determine the various components of this mish-mash of ground cover is a fruitless quest. The main lesson to take away from such photos is that a variety of textures, colors, and heights is going to be needed if we are to approximate the overall look of such scenes (**6-13**).

Photo **6-14** shows how ground cover can be built up with the new 6mm-long static grass material now available from several suppliers. It's topped off with bits of ground foam.

In the May 2006 *Model Railroader*, Peter Ross showed how he "stacked" shorter layers of electrostatic grass to achieve a spectacular tall-grass look (**6-15**).

Modeling tall grass using Silflor mat material was covered by René Gourley in the February 2001 MR. Bernard

Kempinski of Alkem Scale Models uses craft fur to create grass and wheat fields (**8-12**). His advice is to be sure to use the coarse fur, as it's thicker and denser and looks great when trimmed to scale length.

Harold Minkwitz prefers Joann Fabrics Grizzly Fur to model the high, golden grasses of California, as he explained in March 2005 RMC.

A wide variety of grass mats are available to replicate the look of everything from a cow pasture (MiniNature) to a mowed lawn or golf course. To avoid the pool-table look, it pays to add some vertical variations to the surface. Building a slight roll into the landform, as Bernie did with his wheat field, or locating a lawn or field on a slight slope, will create needed texture.

Weed fibers like those made by Woodland Scenics are easy to stick into a foam scenery base using Micro-Mark's U-shaped tool (**6-16**). Try to avoid a flat-topped, well-groomed look, and insert several weed clumps close together rather than scattering them around.

Prairie Tufts and Grass Pathways from MiniNature and Verlinden flocking make it easier to create a weed-grown siding or building foundation (**6-17**). Just be sure not to overdo it to the point that cars derail or engines stall.

Custom layout builder Lance Mindheim likes to use a concrete or beige color in an airbrush, sprayed at a low angle, to vary the color of commercial grass mats.

Bill Henderson's article, "Landscaping by guess (and by gosh)," in the August 1998 MR, is an excellent overview of a wide variety of scenery techniques.

The more, the merrier
The best advice comes from Dave Frary and Bob Hayden, who urge modelers to have samples of a wide selection of natural and commercial scenic products close at hand when adding ground cover to one's model railroad. It's the variety, not any one specific product, that will win the day.

6-14
Bill Darnaby

Bill Darnaby used the new 6mm electrostatic grass to make this convincing field for his HO scale Maumee Route. Bits of ground foam were dusted on to create a weedy look.

6-15
Peter Ross

Peter Ross used Woodland Scenics static grass flock in Noch's Grass-Master to apply not one but two layers of grass, as he explained in May 2006 MR.

Micro-Mark makes a U-shaped tool that's handy for inserting Woodland Scenics tall grasses into a foam scenery base. Thick clumps look better than scattered applications.
6-16

6-17
Bill Schneider

Bill Schneider's diorama features a Branchline Trains kit to depict Munns station on the New York, Ontario & Western. He modeled the weed-grown track by dumping Woodland Scenics foam and cinders in place, soaking it with WS's Scenic Cement, and sprinkling it with Verlinden flocking made for military dioramas. The GE 44-tonner belongs to Mal Houck.

Jim Six

CHAPTER SEVEN

Seasons – All five of 'em

Tom Johnson's Conrail-era HO layout is set in the height of autumn but still manages to avoid a circus-train riot of colors. Note the "frame" of green hues plus the mix of yellows, oranges, and browns on the maples. Moreover, most trees exhibit a variety of leaf colors. Note, too, the typical white paint on the frame building, as discussed in chapter 3.

Most model railroads depict a sunny summer day, while others bask in spectacular fall color (**7-1**). But modelers looking for a new challenge and a unique setting for their modeling endeavors are anchoring their railroads in the late autumn, winter, or early spring, which in New England includes the "fifth" or mud season. This is understandable, as trees and ground cover are typically among the first aspects of a model railroad's scenery to catch the eye. In fact, unless we use some discretion, spectacular scenery may prove to be too much of an attraction, and hence a distraction, much like a discordant squeak from the wind section of an orchestra.

These three photos show what a major role green plays even at the height of fall colors. The Lehigh & Hudson River scene near Roe's Orchard in New York in 1971 (top) shows the drama added with strong shadows, which can be modeled with dark foliage and black paint. The shot of Delaware & Hudson PAs on a 1973 fan trip (left) shows how a dab of color in the weeds can set the tone. Aspens along the Cumbres & Toltec in September (right) are set off by green leaves and evergreens.

7-2

Everything in moderation

At its best, our hobby is about modeling railroading, not just building railroad models. The action, the way we use our models, is as important as the models themselves. The scenery is there to provide a setting for this action, not to upstage it.

Another caveat: No matter how good the scenery, a railroad that is eye-catching to look at but offers little in the way of realistic operation soon gathers a lot of dust. When visiting a model railroad that was not designed with realistic operation as a primary goal, too often it's apparent that the railroad hasn't turned a wheel since the last visitor stopped by. The number-

conscious engineer in me whispers that such model railroads represent an under-utilization of time, money, and talent.

That isn't to say that I haven't been very impressed with some layouts that were built primarily or even exclusively as scenic tours de force. The artist in some of us cannot, and should not, be denied. However, I still recommend approaching scenery as a setting for the action up on the high iron, not as an end in itself. One of the most important attributes of a model railroad is that, unlike even the very best museum diorama, it can actually operate as realistically as it appears.

Planning ahead

The season we choose as a setting for our model railroads may play an important role in telling a story about what the railroad does for a living. Heavy coal traffic in the summer and fall foreshadows the winter heating season. Iron-ore boats have to get to rail connections across the Great Lakes before they freeze solid. Heavy grain traffic follows the autumn harvest. Back in the steam-to-diesel transition era, new automobiles were rushed to dealer showrooms for the traditional debuts in August and September.

Maintaining air pressure in leaking brake lines is also more difficult in the winter, so trains may be shorter or have

Mike Confalone models New England's "fifth" or mud season in the spring using techniques described in this sidebar. The photo backdrop does an excellent job of extending the apparent depth of this foot-wide shelf layout.

7-4

Mike's 1 x 11-foot shelf layout is a "warm-up act" for a much larger layout planned for an adjacent room, but it shows how much railroad can be modeled on an extremely narrow shelf.

Modeling the "fifth season"

In northern New England, there's an unofficial season between winter and spring called the "fifth" (or mud) season. Mid-March to mid-April is a time of brown landscapes, bare winter trees, swollen rivers, and a patchwork of dirty snow and ice (**7-3**, **7-4**). Here are some tips on making realistic fifth-season scenery.

Backdrop: On my 1 x 16-foot Hardwick & Woodbury Division, a 300-dots-per-inch (DPI) backdrop made from April photographs of the area I'm modeling dominates the scene. You'll need a good film or digital camera and access to a graphics house to produce the large prints. Experiment with different distances to see what looks best. Cut around the trees on the horizon. The backdrop wall should be a light gray or extremely light blue to simulate a cloudy-bright day.

Ground base: Lou Sassi's "ground goop," a mixture of Celluclay (from Activa Products), Vermiculite, white glue, and brown latex paint

7-3

Both photos: Mike Confalone

– explained in his book *How to Build & Detail Model Railroad Scenes* (Kalmbach) – is a superb all-purpose base that I use to cover everything. It sets up slowly and is a realistic way to represent muddy earth.

Spring grass: In April, matted, dead, yellow spring grass is everywhere. To model it, I go out into a field in early spring and grab fistfuls of the stuff. I cut it into manageable clumps, throw it in a blender with water to puree it, and wring out the excess water. I put it on a cookie sheet, bake it until dried, then spread it strand by strand or in clumps. I make several grades ranging from long to short strands. I put it down in layers, then mist it with wet water and matte medium.

Forest floor: The leafless trees make modeling the forest floor important. I borrowed Lou Sassi's idea of using real leaves. I blend them, run them through a strainer, then shake them onto the forest floor, also adding broken branches and real rocks.

Gravel road: The back-country gravel road and wood plank grade crossing and crossbucks add a northern New England character to the scene. The road base is Durham's Water Putty, given a loose coating of gray-colored grout (found in craft stores) mixed with rough ballast. This mixture is brushed on and in most places left loose to simulate a well-worn gravel road. The crossing is made from wood strips, stained and weathered. The crossbuck is a length of code 55 rail with a wood sign painted white. The lettering is from my ink-jet printer.

Dirty snow: No early spring scene would be complete without piles of dirty snow. The "snow" is Diamond Dust (found in craft stores), ultra-fine crystals that sparkle like ice. I mix it with gloss medium to form a paste, which is spread and piled here and there. For the dirtiest snow, I drizzle burnt umber and black oil paints, thinned way down with turpentine, over the snow piles. For fresher snow, a bit of baking soda is sprinkled on top of the pile.

Barren trees: I made the bare deciduous trees from a low-lying bush found in New England; I believe it's wild blueberry. Using this base armature would be fine, but I usually add some fine branches using Gypsofilia, attached with hot glue. It's similar to the method Paul Dolkos uses. Grayish-brown spray paint finishes them off.

I also harvested a large batch of incredible trees from the roots of a huge maple tree at my brother-in-law's home in Connecticut. The tree had recently fallen and the stump was exposed. I just happened to walk by it and catch a glimpse of these amazing scale trees sticking out the end. It pays to keep your eyes open and a pair of clippers handy.

The other significant trees are the evergreens, which are up against the backdrop in just about every photo. They approximate "Christmas trees."

The evergreens are made with Scenic Express Supertrees. The armature is painted black or dark gray. While the paint is wet, I used a Noch Gras-Master to dispense dark green electrostatic grass "needles" onto the wet paint. The effect is quite amazing, with just about all needles sticking out instead of laying flat. A bit of pump hair spray secures the needles later. This technique can be used for pine or any other kind of evergreen. As you can see, they blend in effectively.

– Mike Confalone

extra motive power to battle the snows accumulating up on mountains or packed in prairie cuts. Extra passenger trains may be needed to take vacationers to summer resorts. Desired traffic levels and types may therefore weigh more heavily in the decision as to which season to model than sheer scenic beauty.

As I was planning the operation of my new HO railroad, which is set in central Indiana and Illinois in the mid-1950s, I needed to know which time or times of the year were the busiest on the full-size railroad I'm modeling. Most evidence pointed to the fall grain rush when corn, soybeans, and wheat were being loaded into boxcars (this was before big covered hoppers were commonly used as grain haulers) at local grain elevators for transportation to processing plants. So it was traffic density, not scenic grandeur, that pointed me to an autumn setting for the railroad.

Falling for fall

"Fall" is a broadly defined term that covers the period from when little more than a hint of color begins to show on deciduous trees (**7-2, 7-5**) to the appearance of full-blown fall colors (**7-6**) and finally to barren or brown-leafed trees stoically awaiting the first snowfall of winter (**7-7**). Foliage colors peak several weeks earlier in the north than the south. Again, one size does not fit all.

I must confess that when I learned that my railroad needed to depict an autumn setting, I immediately became apprehensive. We have all seen layouts depicting fall foliage modeled as bold blotches of pure yellow, orange, and red trees scattered around the landscape. All too often, the result is a cartoon-like, paint-by-numbers appearance.

So when summer gradually segued into fall last year, I went out with my camera to record the seasonal changes. Careful observation of deciduous trees soon revealed what was wrong with the solid-color-tree approach to fall foliage: Much of the foliage on trees that were, at first glance, yellow or orange was actually still green (**7-5**)! Add in the darker, shadowed areas that we discussed in chapter 6, and many of the

7-5

7-6

Look carefully at the photo above and you'll see that the tree at left isn't all yellow; there is a lot of green and about 20 percent appears black from shadows.

The tree at right qualifies as what Jim Boyd calls an "electric maple," but it still wears a variety of hues from summer green to yellow and orange. And don't forget the shadows.

yellow or orange trees were not primarily yellow or orange at all.

Moreover, only a few of the trees were brilliantly colored – what my good friend Jim Boyd calls "electric maples" (7-6). Most were still green. This observation told me what I had suspected: Green should dominate fall foliage.

To be sure, there is perhaps a week or two during the height of fall color, especially in New England, when the trees are the riotously hued stereotypes that

photographers and painters point to as "typical" fall foliage. But I learned a long time ago that to achieve realism is to model the typical, not the exceptional, and breath-taking color is atypical.

I was not sure whether the traffic levels I wanted my railroad to support occurred before, during, or after the week or two of peak fall color. I therefore decided to approach fall foliage with primarily green trees, bushes and weeds, then add hints of color here and

there until it was clear that the railroad was set in the fall but not during the circus-like acme of leaf-peeper season.

Late autumn

Paul Dolkos set his Boston & Maine HO railroad in November, well after the debate has ended about whether this fall was prettier than last (7-7). He described his modeling techniques in the November 1993 *Model Railroader*, noting then that modeling late fall

7-7

Paul Dolkos

There are only a few barren trees visible in the background, but that's enough to put a chill in the air on Paul Dolkos's HO Boston & Maine. He models November in New England after fall foliage has become a pleasant memory.

7-8

This commercial tree from Scenic Express avoids the artificial one-color problem of some autumn trees.

7-9

Summer foliage can suggest that autumn is just around the corner when hints of fall color are added.

7-10

Spraying high-visibility Day-Glo orange paint on a green tree is another quick way to convert it to a tree representing early autumn.

7-11

Some natural home-decorating materials of various colors were purchased from a craft store and used alongside standard model railroad scenery materials on the author's large-scale project railroad to add some visual texture.

Modeling a snow scene

My N scale Denver & Rio Grande Western layout is set in the Front Range of the Rockies at springtime. Snow at the higher elevations **(7-15)** is common during this time of the year; it can be sunny and warm in Denver and snowing hard at Moffat Tunnel. Modeling snow at the higher elevations (**7-12**, **7-13**, **7-14**) makes the climb to the tunnel seem higher and farther than it really is.

I constructed most of the scenery base in the normal manner before making the snow fly. I used foam-board insulation and Hydrocal plaster rock castings blended together with Sculptamold.

I then masked off the track, roads, creek bed, and the fascia edge. The scenery was thoroughly wetted with a spray mist bottle filled with water and a drop of liquid detergent as a wetting agent. I then sifted dry Hydrocal through an old kitchen sieve to make it "snow." I wetted the surface again to make sure the plaster was soaked. The secret is to make sure to mist on enough water that all of the Hydrocal gets thoroughly wet, but not so wet that it runs all over everything. More plaster can be added using the same steps to increase the depth of the snow.

I "flocked" the trees using this same method, but did it separately on flats made of pieces of foam. This made the trees easy to handle and gave me more control over the amount of snow they acquired, and let me soak them on all sides so the Hydrocal would properly set.

I painted the creek bed a deep bluish-black color and used gloss medium for the water. Snow on top of the boulders in the creek was modeled with a thick artist's gesso, a white primer base that artists apply over a canvas. It works well for touching up snow areas like these rocks since it dries flat, is unaffected by light or water, is elastic, and does not yellow.

For areas of snow that I wanted to build up in deeper layers, I used another product found in art stores: Liquitex Modeling Paste, a matte, opaque preparation of marble dust and polymer emulsion (gesso will crack slightly if applied too thickly). These two products made finishing up the snow effects much easier. Snow effects on other items, such as the road and

7-12

7-13

7-14

Three photos: Mike Danneman

structures, were modeled using sifted Hydrocal and misted water.

The track was covered with Arizona Rock & Minerals snow, which is actually marble dust. To make sure trains still ran smoothly through the snow and ice, I used a homemade wooden template to maintain track clearances and keep the flangeways clear.

Once I got all of the snow nicely positioned and tailored, I glued it down with matte medium in the same way I would glue ballast. Errant globs of marble dust that sometimes crept up the sides of the rails were easily removed with a toothpick. While the marble dust was still wet, I also cleaned out the flangeways with toothpicks. I did this when it was still wet, as the marble-dust snow sets up like concrete when dry.

The new snow on the track dried with a slightly yellow color compared to the stark white Hydrocal snow scenery adjoining it. I easily fixed this with a diluted gesso wash carefully applied over the roadbed. For some added depth to the snow scene, I brushed a very light blue wash in the shadow areas.

Overall, the snow scene ended up being just as sturdy as any other model railroad scenery. As time goes on, it will require only the usual layout maintenance such as vacuuming.

One thing that cropped up that I didn't think about ahead of time was keeping the track clean. With "snow" almost to the top of the rails and in some areas small snow banks lining the track, you can't just swipe a cleaning pad down the rails. The track has to be cleaned much more carefully and usually only one rail at a time. An errant slip with a dirty cleaning pad leaves a black streak on the snow. Blemishes can be fixed by lightly scraping the surface and touching them up with gesso. Now I can empathize with traction modelers with all that street trackage to keep clean!

The snow scene has also added to the operations of the layout. I occasionally call a plow extra to make a run up to Moffat Tunnel and back to keep the railroad open.

— Mike Danneman

7-15
Jim Boyd

A westbound Southern Pacific freight passes above Donner Lake, Calif., in May 1981, before a tunnel rerouted this part of the line under the mountain. This scene could be readily modeled using Mike Danneman's techniques.

would give the railroad a different look.

When doing due diligence about modeling the late autumn, Paul discovered "an interesting palette of browns, tans, reds, and grays as well as hardy greens." Modeling the pre-winter landscape also freed him from having to create "masses of model foliage." The results, by any measure, are spectacular.

Modeling fall foliage

For reasons that should now be apparent, the key to modeling autumn without creating a light show of fake-looking colors is moderation. Trees that are all one bright hue tend to create blotches of color that attract way too much attention to themselves.

Amateur landscape architect Jeff Halloin recommends employing "color echoes" – that is, repeating a handful of colors throughout a scene to tie everything together, reserving bright colors only for focal points. The reds and browns of autumn should echo the oxides and Tuscans of freight cars, for

7-16
John Allen; Bob Hayden collection

John Allen's pioneering Gorre & Daphetid HO railroad included a snow scene at Cold Shoulder high in the Akinbak Mountains. The snow underscored the higher elevations of this part of the layout.

7-17

There's a hint of spring deep in the Appalachians at Hendricks, Md., along the Western Maryland's Elkins line in this May 1973 photo. Everyone loves springtime, but few actually model it.

7-18

Jack Ozanich

Jack Ozanich's freelanced Atlantic Great Eastern is set in early spring in mountainous New England when the dried grasses are matted down from snow cover and leaves have yet to reappear. Most of the railroad is built on a narrow shelf, and Jack uses a simple painted backdrop with tree shapes stamped onto it to extend the background. Jack took this photo using lighting that simulates a gray, overcast day.

example. "Yellows and whites can't be isolated as single trees," Jeff observes, "because they will make the scene look 'spotty.' But they can be combined as a group to make something stand out. Deep greens and gray bare trees are also important, as they give your eye a place to rest as it moves through a scene – not everything should compete for our attention equally."

Several manufacturers make trees that are not solely coated with one garish color of "leaves," among them Accurate Dimensionals, Grand Central Gems, and Scenic Express (**7-8**). Placing them among a stand of still-green trees is an easy way to see whether they complement or detract from the desired big picture. Adding a hint of color to a green tree is easy to do with ground foam or other types of leaf material (**7-9**). A dusting of bright "Day-Glo" yellow or orange from a spray can onto a green tree also works very well (**7-10**).

Most larger art and craft supply stores sell various colors of dried plant material that can be useful to add hints of color here and there to the tangle of brush and weeds that typically populate the borders of a railroad's right-of-way (**7-11**). Bundles of white-bark branches sold by gardening stores as decorations make good armatures for birch and sycamore trees or fallen tree trunks.

The goal is not to model specific foliage and ground cover, but rather to create a busy texture of colors and shapes that looks unkempt and hence more natural. In this case, neatness really doesn't count.

Last, don't forget to model fallen leaves. Lance Mindheim suggests using Model Power sawdust.

Snow season

Modeling the snows of winter (**7-15**) or early spring is comparatively rare. The earliest example that comes to mind was a snow scene near the summit of John Allen's fabled Gorre & Daphetid in the town of Cold Shoulder (**7-16**).

Snow-scene pioneer Brian Holtz described his snow-modeling techniques in the January 1980 *Model Railroader*.

More recently, artist Mike Danneman has devoted substantial

portions of his modest-size N scale Rio Grande layout to spectacular depictions of winter shows in the Rockies. See Mike's "Modeling a snow scene" on pages 74-75. For more photos and information, check the January/February 2007 issue of *N Scale Railroading*.

Need "snowed-on" trees? Check www.photoqualitytrees.com. They also sell yellow aspens and other types of trees made from natural California foliage.

I can't imagine that packing snow between the rails and around the switch points in turnouts is easy, but look at the bright side: There are no grasses, weeds, leaves, or crops to worry about!

An early spring

When I think of spring, chirping birds and a green patina of young leaves beginning to show on tree branches (**7-17**) come to mind. Weather can range from cloudy days and rain to late snows, especially in the north, to T-shirt temperatures. Several local railfans used to drive south from New Jersey to the central Appalachians each spring, as there was just enough foliage to add some texture to the barren winter trees, but there were few bugs or snakes looking for something to chomp on.

Jack Ozanich, a recently retired railroad engineer, sees spring a little differently. Jack's freelanced Atlantic Great Eastern, an HO railroad set in New England, depicts not late but early spring. The lighting and trees and patches of ice and snow suggest a recent spring thaw, but clearly there's more snow in the forecast (**7-18**).

When trees are bare, you can see farther into the forest. This presents a challenge on narrow shelf layouts like the AGE. Jack had rubber stamps made of various sizes and types of tree trunk and branch structures. He pressed the stamps into various hues of gray and brown paint, then printed the images on the sky backdrop, But before he returned to the pad for more color, he reused the stamp several times, each impression naturally being a bit lighter than the previous one. This achieves a sense of distance and hence of more atmosphere between the viewer and the barren trees.

7-19

Kudzu is a fast-growing vine imported from the Orient, and it will cover everything from trees to poles to bridges in short order. The kudzu appears to be gaining a foothold in this September 1975 view on the C&O's Big Sandy line as it nears Elkhorn City, Ky., and its end-to-end connection with the Clinchfield.

In the 2007 *Model Railroad Planning*, Mike Confalone described a section of his HO railroad, which depicts the "fifth season" of a New England spring. He shares many of his innovative modeling techniques on pages 70-71.

A short summer

Countless magazine and book pages have already been devoted to discussions of summertime trees and ground cover, including chapter 6, so we'll ignore that popular season here.

But before dismissing summer as playing second fiddle to the fall harvest traffic rush, consider a comment from Ray Breyer: "I avoided the whole problem of depicting fall foliage by setting my late-1940s granger railroad in June and July. I get the tail end of the livestock rush, the beginning of the winter wheat harvest mini-rush, the middle of heavy construction season, and lots of first-cutting hay being shipped around the country." As I said, there's more to modeling scenery than picking pretty scenes to model.

One last comment about summer: kudzu. This fast-growing vine (**7-19**) seems to be something out of a science-fiction movie, covering Southern hillsides and utility poles and trees with equal aplomb. (How do you plant kudzu? Throw it and run!) Those mod-

eling summertime in the South in recent decades will need to depict kudzu as a thick coating of large, deep-green leaves.

Too many choices?

As model railroaders, we've never had so many excellent choices as we acquire locomotives and cars, structures, detailing parts and scratchbuilding supplies, and, of course, scenery materials. It pays to make some tough decisions up front so that our hobby dollar investments can be narrowly focused on specific goals.

That reinforces my earlier remarks about researching and then picking a season that complements our railroads' operational objectives. You may find, as I did, that heavy traffic levels are directly related to the end of the growing season or the beginning of the heating season. You may find that the railroad you're modeling was especially busy for a few months of only one specific year owing to frequent shipments of construction materials for a new pipeline or highway project.

If so, be sure the scenery treatment of your railroad makes it easier for the audience – family member, visitor, crew member, video viewer, or magazine reader – to see the big picture that you want to convey.

8-1

Ken Patterson

CHAPTER EIGHT

Crops and natural resources

When Ken Patterson was looking for a way to make cornstalks in quantity before Busch introduced a commercial version, he happened to heat up branches from an artificial Christmas trees with a heat gun and discovered they resembled what he was seeking. Much of his success was in the application, however, as the farm equipment and field plus the railroad keep your eyes from dwelling too long on the details of each cornstalk.

I grew up in the Midwest and have always admired the vast sweep of corn (8-1) and wheat fields reaching to infinity in every direction. When my family took a car trip into the Appalachians, my horizons at once broadened and became restricted: I gained a greater appreciation for other landforms, but the juncture between sky and land was often found near at hand – apparently just over the next ridge. I spent a quarter century modeling the impressive Appalachians and railroads that served them. But one day I realized that I wanted to model the style of railroading that I had grown up with: high-speed, single-track lines through granger country (8-2). There was a catch, however: I knew how to model forests, but how should I depict agriculture?

Phil Horning

Nickel Plate S-3 Berkshire 770 accelerates through the cornfields near the NKP's hub yard at Bellevue, Ohio, on July 14, 1957. The elevated right-of-way would make it easy to hide the end of the road at a sky backdrop a few inches behind the track.

The shelf advantage

Fortunately, I was not the only one, or even one of the first, to model the grain belt. Decades ago, the Midwest Railroad Modelers club (May 1987 *Model Railroader*) built a highly linear, narrow shelf layout in Batavia, Ill., that proved very successful. Bill Darnaby was a member of that club, and his Maumee Route home layout was built on its lessons. My new layout, which depicts part of the Nickel Plate's St. Louis line, is in turn patterned after lessons derived from the Maumee.

If I were going to follow their examples by setting my new railroad in the central plains, I was going to have to learn how to model crops in fields. Two things make this easier than in the past.

First, there are now sheets of grass material to simulate everything from spring grasses dotted with colorful flowers to golden wheat fields. And you can buy rows of HO corn, haystacks and bales, vegetable garden plants, fields of flowers, a pumpkin patch, and even plowed fields.

Second, and perhaps most important, the pioneering modeling work done on the former Midwest Railroad Modelers club layout showed that we need not model the acreage in the entire "southwest 40" to simulate a field of crops. (Land was divided into 160-

Modeling coal mining and agriculture isn't an either-or proposition, even in the Appalachian foothills. This recent fan trip scene on the Ohio Central in southeastern Ohio isn't far from the soft-coal tipples that contributed to former owner Wheeling & Lake Erie's (later Nickel Plate) bottom line.

acre farms under the Homestead Act of 1862.) Rather, we can model only that swath of land that comprises a railroad's right-of-way. The fascia marks one edge of the railroad's property; a fence and pole line marks the other. Behind it is a sky backdrop, with the fence and tree line hiding the right

angle between the layout surface and the sky.

This achievement precluded the need to model endless acres of corn or wheat or soybeans. It also showed that we can achieve the look of flatlands railroading on narrow, space-saving shelves, which in turn facilitate the use

8-4

Santa Fe Ry.

Four brand-new Santa Fe FT A and B units on their first trip east to Chicago have their train of refrigerator cars well in hand as it curves among dipping sedimentary rocks near Cajon Pass.

8-5

Bernard Kempinski

Regional crops help identify the location of a model railroad. These tobacco plants were made from brass etchings by Alkem Scale Models.

8-6

Both photos: Allen Keller

The orderly rows of soybeans on Cliff Powers' Mississippi, Alabama & Gulf (left; also see *Great Model Railroads* 2007) are bits of Woodland Scenics clump foliage over corrugated cardboard.

The crops Allen Keller has modeled on his Bluff City Southern (below) reflect the economy of the late-steam-era Mississippi Valley. His cotton plants (foreground), for example, are chunks of ground foam with dots of white paint applied to each "plant." Even the soil color fits the region.

8-7

of narrow peninsulas to increase main-line runs, an important consideration when modeling one of the Midwestern race-track railroads such as the Wabash or Nickel Plate.

Tips on using commercial products to model various crops are offered in the accompanying sidebars.

Regional crops and resources

It's easy to think of corn and grain as mainstays of the landscape in the Midwest and in central Canada, but crops are also common in the flatter river valleys in the mountains (**8-3**). You may be surprised to learn that 40 percent of West Virginia's land area is farmed (including woods and pastured woodlands). In 1964, most Mountain State farmers raised livestock; then came dairy, poultry, field cropping, general, tobacco, grain, and vegetable. Corn was one of the state's principal agricultural products.

Good climate, a gift of the Pacific Ocean currents, and flat terrain bracketed by coastal mountain ranges have made California's Central Valley a prolific producer of perishable goods. Endless strings of ice and later mechanical refrigerator cars ("reefers") racing eastward from the Golden State (**8-4**) are popular modeling subjects. The Pacific Northwest is known for its apples and Idaho potatoes.

Commercial fruit trees and garden plants are available from a variety of manufacturers. As with corn, we can use only a row or two of trees or plants along the backdrop or fascia. Additional rows of trees can be painted or rubber-stamped on the backdrop.

For those who model the railroading of the South, Alkem Scale Models makes brass etchings that fold into convincing tobacco plants (**8-5**). Nearby storage warehouses underscore the crops grown in the fields. Palm trees (**1-6**) are also available from several suppliers.

Maine-grown potatoes were important to the financial well-being of Down East railroads such as the Bangor & Aroostook. Like any perishable crop, however, they're seasonal, so the time of year depicted on a layout based on the BAR would ideally complement

8-8

Modeling corn

Busch sells plastic cornstalks that have greatly eased the chore of planting cornfields, as Cody Grivno demonstrated in the September 2006 *Model Railroader* (**8-8**). But it takes quite a few of them to model a garden, let alone a field. "Plant" a couple of rows of corn along the front or rear edge of the railroad right-of-way to extend them; use a stiff, almost dry brush to stipple some olive-green paint onto the backdrop behind the cornstalks to add apparent depth to the field.

Corn rows were initially spaced far enough apart to allow room for horse-drawn farm equipment. Spacing was reduced from 40" to 20" as machinery supplanted the oat burners. Crops can't be planted right up to the fence at the end of each row, Mark Plank reminds us, as the horses or tractor have to turn. "Header" rows (4, 8, or 12 rows wide if a 4-row planter was used) were planted across the ends of the field. They'd get run over during cultivating or spraying; some plants wouldn't fully recover.

Alkem Scale Models makes etched-brass cornstalks, which can be twisted into a 3-D shape (www.alkemscalemodels.com). You can use the flat etchings as a master to make a rubber stamp to imprint on the backdrop. Bill Darnaby explained how he used Alkem corn in the June 2003 MR.

Ken Patterson heated branch tips from an artificial Christmas tree (**8-1** and **8-9**) to make an entire cornfield; see the August 1999 *Mainline Modeler* and June 1995

8-9

RailModel Journal for details. Others have used artificial turf with every other row cut out; plastic asparagus fern, sold at art-supply stores (smash and turn the leaves using pliers); and green turf doormats with every other row trimmed down and the tips dusted with yellow ground cover.

To depict the approaching harvest season, the lower portions of the stalk could be painted tan to show that they're beginning to "fire" (**8-10** and **8-11**). Busch now offers a field of harvested corn.

In N scale, Patrick Lana uses sill sealer cut into ⅝" strips with the top edges trimmed at 45 degrees into a point shape. He glues the strips side by side to form rows, each following the natural contours of the land, with the plastic skin peeled off the outer edges to provide texture. Patrick paints the installed corn Apple Barrel Leaf Green, Forest Green, and Folk Art Clover. – *T.K.*

8-10

8-11

8-12

Bernard Kempinski built this N scale layout section using 9" x 12" sections of craft fur cut to scale height using hair clippers and scissors. (Fake fur proved unsatisfactory when trimmed that short.)

Amber waves of grain

Bernard Kempinski used craft fur cut to scale height to model an N scale scene (**8-12**). Harold Minkwitz sprinkles Woodland Scenics fine yellow grass turf on faux fur to make wheat fields (see www. pacificcoastairlinerr.com/fur_grass). Small wheat fields could be made using the new 6"-long electrostatic grass as shown in chapter 6.

Commercial wheat field mats come in sheets. Seamlessly splicing two or more sheets to create a huge expanse of wheat can be done with some care, but narrow strips of wheat between the aisle and right-of-way or right-of-way and backdrop are easier to manage (**8-13**). I used an amber-colored mat to simulate wheat ready to be cut.

Unlike corn, young wheat plants can be run over as a field is cultivated or sprayed without much lasting damage. Winter wheat is planted in the fall and often fertilized in the spring, but the mashed plants usually recover, Mark Plank reports.

The Pendon Museum in the United Kingdom features very large scenes of farming in the 1930s. It's built to 4mm scale, slightly larger than HO's 3.5mm (1:87), with the wheat fields made from hemp rope. After lengths are glued down using white glue, the rope is teased and trimmed.

Ryan Laroche points out that in Canada there are two grain rushes: in the fall to get grain to Great Lakes ports before the lakes freeze over, and from April to June as grain still sitting in elevators is sold to make room for the fall harvest.

– T.K.

8-13

Busch and other suppliers make various colors and thicknesses of grass mats that can be used for meadows and crops such as wheat.

the peak potato shipping season if traffic levels are important.

Deep scenes

Convenient as it is to keep agricultural scenes narrow to require fewer plants per field, it's also nice to have a bit of room to spread out when modeling a field or garden. This is often easier to accomplish on single-deck model railroads, as supports for the upper deck aren't a factor.

Cliff Powers' narrow, linear layout employs photo backdrops to expand his fields (**8-6**). Allen Keller's Bluff City Southern is a linear railroad, but he has used the area inside the curves at the ends of peninsulas to model wide expanses of various crops such as cotton and tobacco appropriate to the railroads between Memphis, Tenn. ("Bluff City" is a nickname for Memphis), and Birmingham, Ala. (**8-7**).

Trees are considered a cash crop and are ground or leached into pulp to make paper or cut into lumber. The Piedmont area between the Appalachians and the Atlantic in the southeastern U.S. is an ideal home for vast forests of pine trees, hence the paper and furniture industries located there.

Natural resources

Money-making opportunities for railroads are found under as well atop the ground. In the Southeast are deposits of clay, not the least of which is kaolin, a white clay used to coat paper and make porcelain fixtures. Deposits of mica, used in paint and insulators, are found along the former Clinchfield. The Sunshine State hosts fruit trees as well as extensive deposits of phosphate, which makes good fertilizer and creates thousand of annual carloadings.

I documented the importance of hard and soft coal to the economic health of railroads in my book *The Model Railroader's Guide to Coal Railroading* (Kalmbach, 2006). Railroads also play a key role in the transportation of petroleum products to market. What began in western Pennsylvania has since spread across the country, and imported oil has to be refined and transported by railroads, trucks, and pipelines.

The *Keystone Coal Industry Manual,* published by McGraw-Hill, lists not only coal mine locations and names but also Portland cement plant locations by state (also check www.cement.org/index.asp). The 1972 edition lists plants in 41 of the 48 contiguous states. Note that cement is an ingredient of concrete, so a concrete batch plant would require not only a source of cement but also aggregate. State departments of natural resources may have background information of such large industries online; Iowa's cement industry, geology, and other natural resources, for example, are documented at www.igsb.uiowa.edu/browse/cement/cement.htm.

Another regional industry was glass making. Eric Hansmann documented examples of this industry in West Virginia in the 2001 *Model Railroad Planning.* Henry Freeman showed how he built a PPG glass factory on a 4 x 8 sheet of plywood as a plug-in section to his basement-size B&O layout in January 2003 MR.

Wallboard is a key product in construction. It's made from gypsum, and where better to find that mineral than Gypsum, Colo., near Glenwood Springs on the UP (formerly Rio Grande)? But cities and towns named Gypsum are also located on railroads in California, Iowa, Kansas, Ohio, Texas, and Newfoundland.

Scenery earning its keep

The learning point here is that, while railroads were occasionally built to haul tourists to national parks or commuters to work, most were built to haul coal, oil, timber and lumber, livestock, grain, perishables, or other major products and natural resources to market. The modeler who wants his or her railroad to operate as realistically as it looks, and to look like a railroad that earns its living in well-defined ways, needs to identify its main sources of income and ensure that the observer can easily determine that traffic base. More than simply being eye-catching, scenery therefore has a major role to play. Modeling specific crops and natural resources may be critical to achieving this goal.

Soybean rows

Soybeans became a major cash crop in the 1940s. In fact, the largest industry on my railroad is the Swift (later A.E. Staley and now ADM) bean processing plant in Frankfort, Ind., which I described in detail in *How to Build Realistic Layouts, Industries You Can Model,* a special 2007 issue of *Model Railroader.*

Soybeans used to be planted using a corn planter, which typically had 33" spacing (**8-14**). Today, they grow as a lush, green carpet (**8-15**) and are "sweep" harvested with a combine/harvester.

As fall approaches, tinges of bright yellow appear (**8-16**), and the plants finally become solid yellow. The leaves then turn brown and fall off, leaving rather scruffy and wiry looking stalks (**8-17**).

I created rows of soybeans with Woodland Scenics dark-green clump foliage (**8-18**). For variety, I dusted some bean fields with bright yellow foam to show them later in the growing cycle. I have seen adjacent fields of green, yellow-tinted green, all-yellow, and leafless brown soybeans all at one time, apparently reflecting slightly different planting dates.

N scale modelers report various types of corduroy make good soybean plants. Patrick Lana recommends gluing rows of forest-green chenille yarn over a corrugated cardstock base with rows spaced about ¼" apart to simulate the typical 30" to 36" row spacing. Last, he dusts on a light coat of Woodland Scenics green grass.

Crops tend to be planted more densely now than in the steam or early diesel era. The actual spacing is not critical; rather, use a model tractor as a gauge to ensure proper spacing. – *T.K.*

9-1

Allen McClelland

CHAPTER NINE

Backdrops

The hand-painted backdrop on Allen McClelland's original Virginian & Ohio HO layout greatly expanded the scope of the scenery, as shown in this view just west of Fullerton, Va. This scene dates to 1984, when uncoated lichen was still used for forests, but Allen's scenery clearly evoked the central Appalachians.

There are two schools of thought about effective backdrops: Keep them "loose" and somewhat abstract to avoid having them compete with the railroad for the viewer's attention; or, conversely, make them as realistic as possible, like a photograph, so their lack of realism doesn't detract from the realism of the overall scene (**9-1**). Since there is a fine line between artistically abstract and downright sloppy landscape painting, I lean toward the photographic school in this regard, as I feel qualified to judge the quality of a photomural. Fortunately, technology has kept pace with our needs, as today one has a choice between buying or making one's own photographic backdrops in long strips.

Perspective

It's almost impossible to paint a two-dimensional image on a backdrop that looks good when viewed at any angle in relationship to 3-D foreground scenery, but there are several tips to keep in mind as you select images for or paint images on a backdrop.

Note the barn complex in the two photos that comprise 9-2. These two photos were taken with the camera moved 90 degrees between exposures, yet the barn looks three-dimensionally correct to a reasonable degree in both views. That's because it was originally photographed from one corner rather than flat on. Conversely, a flat-on image will look correct only when the viewer or camera is precisely at right angles to the backdrop. The lesson, with some important exceptions (**9-12**), is to use images as seen from an angle.

Also consider the height of the scene relative to your eyes. Backdrops tend to look better when viewed from near eye level, but if all or portions of your layout are set well below eye level, images of structures and roads on the backdrop might be chosen to match the "helicopter" view. They'll look odd in scale-height eye-level photos, however.

Using smaller scale buildings and trees can achieve a sense of distance without creating perspective concerns if you can control the viewing height (**9-3**) – the higher, the better – or at least compose photos of the scene with care.

Hiding the backdrop interface

Hiding the abrupt joint between flat scenery and vertical backdrop is tricky. In *Model Railroad Planning* 2002, Paul Dolkos suggested creating a small rise and then a dip in the road just in front of the backdrop (**9-4**) so that the actual road-backdrop interface is hidden behind the hill. He also recommended doing essentially the same thing when a sloped structure roof is truncated by the backdrop.

Another way to disguise this joint is to create an upwardly curved fillet between the horizontal "ground" and the vertical backdrop where a road or field or even a body of water runs into the "sky." The rounded ramp eases the harsh intersection that would otherwise be clearly evident. I've seen layouts where such fillets nicely disguised the usually distracting intersection of the flat surface of a lake or harbor area that extended several feet along the backdrop.

It's a bit harder to hide this joint when the railroad penetrates the backdrop. One oft-used approach is an overhead highway or railroad bridge (**9-5, 9-13**), but consider lighting the area behind the backdrop.

Clouding up the picture

I asked at the beginning of this book why anyone would attempt to fill an entire basement or garage with scenery without having the same sort of information we expect to have when scratchbuilding a freight car or structure: plans, photos, and so on. The same applies to clouds. With a little understanding about how they come to be, we can more easily make informed choices about how to replicate them on our backdrops.

Let's start with cumulus clouds – those puffy, fair-weather "heap" clouds. As any sailplane (glider) pilot can tell you, they often form when the sun heats up areas of the ground – a paved parking lot, a plowed field, a rock outcropping. The heated air, called a thermal, spirals upward into unstable air. The moist warm air rises and expands until it's cool enough for visible condensation to form into a cloud (**9-6**).

Since condensation happens suddenly at a specific altitude (which edges higher as the day goes on and the air mass heats up), cumulus clouds have flat (actually, upwardly domed) bottoms

Photos taken 90 degrees apart of Duquesne Quick-Copy's Realistic Backdrops farm scene (realisticbackdrops.com) show that backdrop images photographed or painted at an angle look reasonably correct from almost any angle. "Flat" images or even structure flats often look odd when viewed or photographed at other than a right angle.

9-2

Paul Dolkos used N scale structures in the background of his HO Boston & Maine layout to suggest much greater distances than he actually had to work with. The depot is full HO scale.

and billowing tops formed by the up-rush of warm air. If there's a lot of energy and unstable air available, they may continue to bloom upward until they form cumulonimbus (thunderstorm) clouds 10,000 feet, 20,000 feet, 30,000 feet, or even higher.

Cumulus clouds may align into "streets" (**9-7**) where long lines of them are paralleled by bands of blue sky as the air that was uplifted in thermals cools and tumbles out the top of the clouds and cycles back to ground level. You may also observe that those pleasant little cumulus clouds soon overstay their welcome and cover the entire sky,

shutting off the sun (**9-8**). This is a form of cloud suicide, as the lack of ground heating ends cloud production. Often, the sky clears again, and cumulus clouds form anew.

The key to representing cumulus clouds on a backdrop is to create a rounded, puffy top and a relatively flat bottom. This is easily accomplished with stencils sold by New London Industries (www.newlondonindustries. com). Just tape the stencils to the blue-painted backdrop and dust on a light coating of flat white spray paint (**9-9**). To suggest one cloud in front of another, move the stencil down and

spray on another light coat of white. If you want to simulate the darker underside of thicker clouds, use a flatter portion of the stencil turned upside down and spray on a thin medium-gray coat.

You'll find that overspray tends to lighten the sky below the clouds. I therefore don't bother painting in a lighter horizon band. But in areas with little or no cloud cover, you may find that extending full-rich sky blue paint right down to ground or tree level looks a bit cartoonish. If so, paint the backdrop white at the horizon and, while still wet, blend in more and more sky blue as you gain height.

I've also seen cumulus clouds painted using rounded, coarse sponges fastened to the end of a stick or dabbed on by hand. I think the stencils are easier and more effective, as it's easier to create a cloud's wispy appearance.

Observing the sky on a day that you'd like to model will pay dividends, as clouds help to convey a certain mood. If you're looking for the flat lighting of a gloomy spring or fall day, perhaps a hazy sky or a layer of stratus clouds will be more useful. This is the layer that forms as a warm front slides up and over colder, denser air as the front approaches. Harry Bilger's hand-painted backdrop suggests a layer of clouds coming over the distant ridge (**9-10**); the skies may not be so sunny tomorrow.

High altostratus and cirrus clouds (**9-11**) can add some texture to that blue-sky area above your merry bands of cumulus clouds. Wave clouds can be seen as long bands that remain stationary downstream of mountain ridges; they don't drift with the wind. They're called lenticular clouds – "lennies," for short – as a cross-section looks like a lens or airplane wing's airfoil. Look carefully and you may see primary, secondary, tertiary, or even more lenticular formations. These could be modeled with linear streaking of white spray paint.

Those who model east-west railroads often try to keep east to the right, which permits valance lighting to simulate sunlight on the south side of objects. When this can't be done, clouds can suggest the viewer is look-

Another scene on Paul Dolkos' HO layout shows how a slight rise in the road hides the joint with the backdrop. A fillet between the road and backdrop can also disguise the joint.

The photo above shows a simple ridge line painted on the backdrop on the author's Allegheny Midland, but the illusion of distance is compromised by the dark shadows under the road overpass. The two photos at right show the before-and-after appearance of a similar scene on Paul Dolkos' B&M layout after basic scenery and lighting were added beyond the backdrop.

Paul Dolkos

Paul Dolkos   **9-5**

ing south if they are painted mostly in shades of gray to represent the shadow (north) side of the clouds.

Photo backdrops

The effectiveness of a photographic backdrop is clearly evidenced on Doug Tagsold's HO and On3 Rio Grande layouts (**9-12**).

"I took the depot photos while attending the NMRA National Convention in Denver in 1991," Doug recalls. "This was before digital cameras, so I used print film and had 20" x 30" poster-size enlargements

made. The scene comprises nine separate prints overlapped a few inches on each end, which provided 17 feet of accurate downtown-Denver backdrop."

Doug mounted the prints directly onto the blue sky painted drywall backdrop. Amazingly, there are no freestanding flats. The only parts of the station he actually modeled in 3-D are the station platforms and the roofs over them. Doug stood a few hundred yards away from the station across an open abandoned rail yard to take these photos without obstructions. Since 1991, the area has been developed as Coors

Field (the Colorado Rockies' baseball park), so getting the photos today without obstructions would be difficult. However, with today's digital imaging, removing unwanted items from a scene or correcting perspective is practical. For more information, see the June 1995 MR.

Commercial sources for photo backdrops include Railroad Graphics CD (www.geocities.com/larcproducts/), which sells CDs loaded with a variety of backdrop scenes, and SceniKing by BPH Enterprises (www.sceniking.com).

Puffy cumulus clouds crowd the Midwestern sky. These form at the altitude where water vapor condenses as thermals rise and cool; their bases move higher as the day warms up. Cumulus bottoms are thus somewhat flat; their tops billow upward.

9-6

9-7

9-8

Cumulus clouds may form into "streets" – long, almost unbroken chains of clouds with lanes of blue sky between them. Individually or in streets, fair-weather cumulus clouds tend to brighten one's day and can be used to achieve a desired mood on a layout.

Cumulus clouds may over-develop and kill themselves off by shutting off the sun's heating of the ground, or they may develop into rain clouds (the vertical streaking between the clouds and the ground is rain) or even monster cumulonimbus thunderstorm clouds.

New London Industries makes several handy cloud stencils that do a good job of representing cumulus clouds. These photos show the clouds after a single dusting of white paint (top left), with a second, lower cloud line made by moving the stencil down (top right).

9-9

9-10

Harry Bilger

Instead of smoothly blending white and sky-blue paint from the horizon upward, Harry Bilger deliberately introduced streaks that nicely mimic an overcast day (above). The use of blues for the rear-most mountain range conveys a sense of distance in his HO railroad, which is set in Idaho.

A Union Pacific coal train heads eastbound near Lusk, Wyo., in October 2003 (right). The wispy high clouds could be easily simulated with streaks of white paint, much as Harry Bilger did.

9-11

Jeff Wilson

Both photos: Doug Tagsold

Doug Tagsold made excellent use of photos on both his HO (above) and On3 (left) layouts. Denver Union Station and the cityscape are simply photos mounted on the backdrop; only the station platform sheds are 3-D. The distinctive dipping sediments in the rock bluff north of Durango is another series of photos glued to the backdrop to add considerable depth to the scene.

9-12

Ultraviolet rays from sunshine and fluorescent lighting may fade photographs and some printing inks over time. In the 25 years that I spent building, scenicking, and operating the Allegheny Midland, however, I could find no clear evidence of color fading from the cool-white fluorescent lighting that I used, but I didn't use printed or photographic backdrops. Libraries and museums that are concerned about possible harm from UVs often insert fluorescent tubes in clear-plastic sleeves. I understand that the plastic diffusing "lenses" or covers found on some fluorescent fixtures also screen out UVs.

There are inks available for computer printers that are marketed as UV-resistant. Before covering a large area with printed or photographic backdrops, it's prudent to ask about possible fading from ultraviolet exposure.

Tracing photos

Another way to use photos when making backdrops is to project them onto your backdrop, trace around the desired image, and then refer to the photos as you paint in the details. Avoid full-strength colors, or the background will look too close and inky.

Check with your local college or art supply store, as they may offer courses in landscape painting that would add a new dimension to your hobby. But even simple free-hand landforms can add a sense of depth to a scene (**9-5**).

Background texture

I mentioned the cloud stencils from New London Industries. They also make stencils that allow you to paint ghostly "silhouettes" of background mountains, tree lines, hills, and cityscapes.

You might try adding some texture to background tree lines and verdant ridges by gluing ground foam to the backdrop. The innovative New England, Berkshire & Western HO club layout at Rensselaer Polytechnic Institute in Troy, N.Y., uses foam-covered cutouts of background hills. Dick Flock glues foam directly to photos of trees pasted to the backdrop to achieve a 3-D effect.

Jack Ozanich added depth to his shelf layout by using rubber-stamped trees, as we discussed in chapter 7.

And don't overlook mirrors as ways to extend a scene. On his original Virginian & Ohio, Allen McClelland extended a yard scene through a wall by running the tracks under an overpass and putting a large mirror over the bridge (**9-13**). Jim Hertzog used small mirrors to extend several scenes.

Materials

I used tempered hardboard (the dark brown stuff) for backdrops, sealed on both sides. I found out the hard way that latex paint does not seal anything, as it "breathes." Use a sealer.

I notched the top edge of the lower-deck backdrop (**9-14**) to fit around the L-shaped, stamped-metal shelf brackets that support the upper deck. These notches make it difficult to slide the backdrop material in place on curves, so it's doubly important to put up the backdrop between decks on a multi-deck layout prior to installing roadbed, track, or scenery.

Many modelers are now using sheet styrene for backdrops. Squadron Green putty is a good choice to smooth the end joints between adjacent sheets.

Others use rolls of linoleum or aluminum sheet metal. Putting it up is typically a two- or three-person job.

If the backdrop will be hard to reach once the benchwork is erected, you may want to plan each scenic feature well enough to know in advance how the backdrop should appear and how high above the floor it needs to begin and then paint it in advance.

Modeling as artwork

When you think about it, building scale models is indeed an art form. It's perhaps most apparent when we contemplate ways to create a realistic backdrop for our model railroads, but everything we do from building and weathering a boxcar to "planting" a row of trees is an expression of our artistic skills.

Just as not all paintings or sculptures are as successful as the artist had hoped, some of our scenery and backdrop-making efforts will come up short of the mark. That's to be expected. Regroup and move on; you'll do better next time.

Allen McClelland

The V&O's west-end yard at Jimtown, Va., was extended into hidden staging through a hole in the wall under an overpass (above). A mirror above the bridge and lighting and minimal scenery beyond the wall extended the scene. Jim Hertzog used a much smaller mirror to extend a wye track; there are only two hoppers in this scene on his HO Reading layout (right).

Jim Hertzog **9-13**

9-14

The author notched the lower deck's sky backdrop (⅛" hardboard) to fit around the shelf brackets used to support the upper deck. The notches are invisible from normal viewing angles. Installing the backdrop after the roadbed and track were in place proved very difficult.

10-1

CHAPTER TEN

Putting your railroad on the map

For a model railroad, especially one that's freelanced, to find its place in the global transportation network, it helps to locate it on a map. The author used U.S. Geological Survey quadrangle topographic maps to locate the Allegheny Midland's path from southeastern Ohio to western Virginia.

When I planned the Allegheny Midland back in the early 1970s, one of the first steps was to decide what it should do for a living. I greatly admired my friend Allen McClelland's coal-hauling Virginian & Ohio RR, so I created a connection between the V&O's Afton Division and my hometown favorite, the Nickel Plate Road. I grew up along the NKP in Indiana, but I focused instead on the NKP's former Wheeling & Lake Erie line into the coalfields of south-eastern Ohio. All I had to do was "connect the dots" – Dillonvale, Ohio, on the NKP, and the west end of the V&O's Afton Division in western Virginia. That meant putting the railroad, both literally and figuratively, on the map (10-1).

Steve King

This section of a U.S. Geological Survey topographic quadrangle map shows a wide spot along the Cheat River for a yard at mythical Midland (just north of the Western Maryland connection at Parsons, W.Va.).

The author is shown inspecting the same wide, flat area shown in 10-2 when he, Allen McClelland, Jim Boyd, and Steve King toured the AM's mythical right-of-way in 1976.

Plotting the AM's route

I once worked for a land-surveying firm and knew how to read contour (topographic) maps. I also used U.S. Geological Survey quadrangle maps to find long-abandoned railroad lines (10-6). Why not employ them to locate a route for a freelanced model railroad?

I purchased USGS quads for the proposed route from southeastern Ohio through eastern West Virginia into Virginia. With only a modest amount of head scratching, I found a route through the Alleghenies (10-2). Doing that was a lot of fun and a key step toward the creation of a plausible freelanced railroad and scenery that "looked the part."

Later on, at the visitors' center in Cumberland Gap National Park, I found Hubbard vacuum-formed plastic topographic maps for the south end of the AM (10-4; try McCarthy's Geographics in Pittstown, Pa., www. mcmaps.com). These made a plausible route through the central Appalachians much easier to spot – just look for the gaps in the ridges and run through them!

The American Association of Petroleum Geologists (P.O. Box 979, Tulsa, OK 74101) sells geological highway maps for 11 different regions of the U.S. (10-5).

Quiz time

I purchased USGS "topo" maps not only for the route traversed by the

freelanced Allegheny Midland but also for the entire Third Subdivision of the NKP's St. Louis line, which I now model, plus several short lines of interest.

To illustrate their usefulness, let's examine one corner of the USGS quadrangle map that shows Veedersburg, Ind. (10-6). How many railroads do you see? Today, you'll find no railroads in Veedersburg. Until just after World War II, however, three railroads went through town. The one that heads roughly north-south through town was part of the Nickel Plate's St. Louis line (Norfolk & Western by the time this map was updated). The east-west line

Vacuum-formed 3-D maps can occasionally be found. This map shows Elkins, W.Va., at upper left. The Allegheny Midland stayed east of Elkins and followed Glady Creek south to a connection to the Western Maryland near Glady at left center.

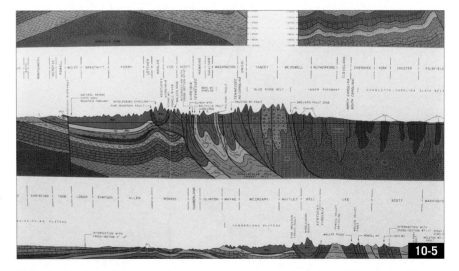

Geological cross-section maps give strong clues about rock formations. This cross-section from a map of the Mid-Atlantic Region, published by the American Association of Petroleum Geologists, shows a vertical fault where slightly dipping sedimentary beds have been shifted upward along today's Kentucky River.

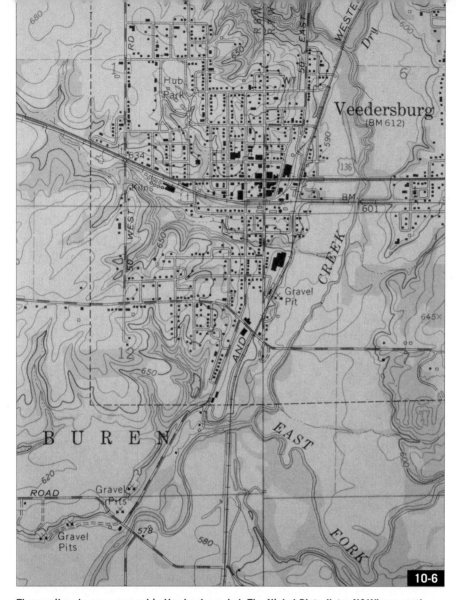

10-6

Three railroads once crossed in Veedersburg, Ind. The Nickel Plate (later N&W) ran north-south through town and crossed the east-west Peoria & Eastern (Penn Central). But can you find topographic evidence of the location of the third railroad? Hint: It roughly paralleled the NKP; look for angled streets in town and bunched contour lines marking a curving fill south of town. See 10-12 for the solution.

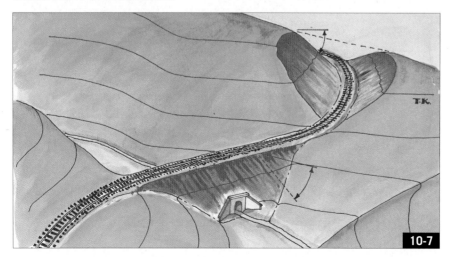

10-7

The author's sketch shows how horizontal contour lines define the shape of natural and man-made landforms. Note how a cut or fill causes the contour lines to bunch up and become parallel, making it easier to trace the path of even an abandoned railroad.

that crossed it was the Peoria & Eastern, part of the New York Central System – by this time, Penn Central.

The "missing" line was the Chicago, Attica & Southern, part of a Chicago & Eastern Illinois branch into the southern Indiana coalfields and originally the Chicago & Indiana Coal Ry., which operated camelbacks! Can you find its route?

Hint: The line came into town from the north parallel to and west of the NKP, crossed the P&E a hundred feet or so west of the NKP, crossed the NKP at the south edge of town, and then crossed Coal Creek just east of the NKP and curved to the southeast as it left town. Look for roads and linear contour lines that don't seem to flow naturally or fit the town grid.

Refer to **10-12** to see the CA&S's path through town.

Reading topographic maps

Figure **10-7** is a drawing of a short stretch of railroad that cuts through two hills and bridges a low valley on a fill. (Civil engineers try hard to balance the amount of material taken out of cuts with that needed to create fills so they don't incur extra costs hauling excavated material away or bringing in fill.) Note how the contour lines for each ten feet of elevation bunch closely together when the slope of the land climbs or falls off abruptly, just as they did on the CA&S fill south of Veedersburg.

The Geological Survey has free index maps of each state showing the names and locations of topographic maps. Locate the prototype or freelanced railroad's right-of-way on an index map to determine which quads to order.

DeLorme (www.delorme.com) publishes handy "Atlas & Gazetteer" books of topographic maps for each state. SPV's Railroad Atlas of North America (www.spv.co.uk), now covering all but three north-central states and some of Canada, is another handy reference when trying to find a railroad – or where a freelanced railroad might fit in.

Using the contour lines

You can accurately model landforms simply by transferring the contour lines from a topographic map onto layers of foam board that represent the intervals (height) between contour lines. If the lines are spaced 10 feet apart, for example, that's about an inch and a half in HO, which can be represented by layers of 1" and ½" blue or pink foam. (Avoid white foam, as it crumbles easily.) Photo 10-8 shows how I used foam to create sufficient depth for the river on a large-scale project railroad.

Sanborn fire-insurance maps

Back in 1866, surveyor D.A. Sanborn devised a system of maps to help fire-insurance underwriters and agents determine risk (10-9). These maps show building base dimensions and give clues as to the structures' construction, height, and orientation. They also show roads and railroads, although the rail lines are often simplified. The maps were updated with overlays pasted atop the original maps, but you may be able to peel back the layers to a particular era.

Many large university and county libraries and some historical societies have copies of local Sanborn maps or even access to a state collection. For example, try www.edrnet.com/sanborn.htm or www.lib.utah.edu/digital/sanborn.

Other types of maps

Railroads often created colorful maps to show potential customers the advantages of locating along their main lines (10-10). We can use such maps to determine what to load in freight cars coming from or headed to a specific region.

Libraries and state government agencies share maps and other documents online. The Indiana State Library, for example, sent DVDs containing county maps to all public libraries; Mark Plank reports they're now online at www.indianamap.org.

Other maps that turn up at railroadiana shows or online point to major cattle ranching and agricultural centers, oil and coal fields, iron ore deposits, and other businesses of interest.

Chris Webster notes that many official state highway and railroad maps are available online; for Illinois, try

Stacks of 2"-thick blue and pink foam board make it easy to build up scenery and cut it to shape river valleys, hills, and cuts and fills. This in-process scene is on a large-scale project layout the author built for *Model Railroader*.

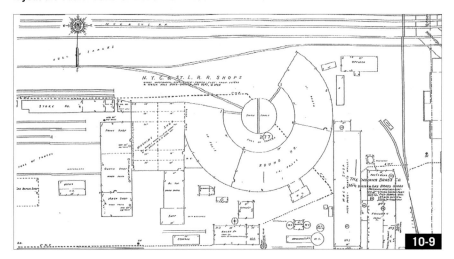

Sanborn fire-insurance maps often provide helpful information about lineside structures, especially if you can find an updated version for the era you're modeling. Many state university library systems have virtually complete sets.

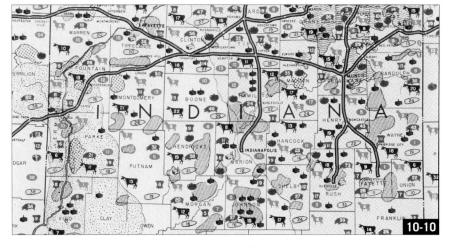

Railroads produced many maps of their own. Some, like this one, were produced to show potential customers natural resources found along various lines. Others were designed to show possible industrial-development areas.

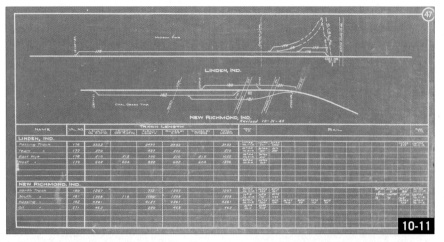

The Nickel Plate track diagram book for the Clover Leaf District included drawings of each town as well as the name of each track, but it can be misleading: This diagram from the 1930s fails to show a short crossover that was added in the 1940s to the passing track west of the wye tracks (at Linden, Ind.) to make running around pickups and setouts easier.

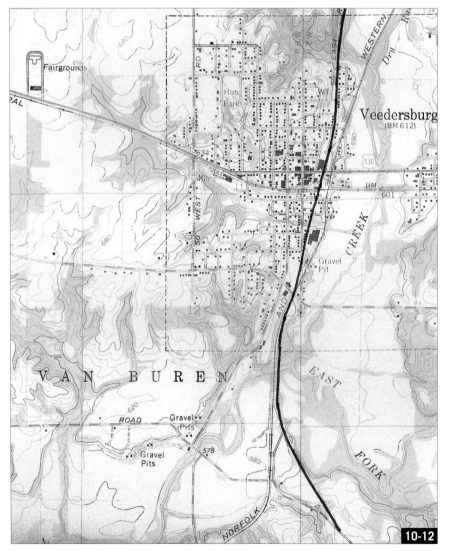

The red line shows the approximate route of the Chicago, Attica & Southern through Veedersburg, Ind. (see 10-6). Note the curving fill south of town that clearly denotes the line's location after it crossed Coal Creek. The line was abandoned just after World War II.

www.dot.state.il.us/officialrailmap.pdf, which shows a recent map. The Washington State Department of Transportation has developed a search engine that checks all state DOT Web sites: www.google.com/coop/cse?cx=0065113 38351663161139%3Acnk1qdckodc.

During the early 20th century, the railroads were asked to account for the value of their holdings and so produced valuation maps. Railroad-specific historical societies often have copies of these important maps.

Railroads often made track diagram books showing the number and name of each track in a city or town as a communication aid (10-11). They are sometimes available from rail historical groups.

Photographs

Aerial photos are available from a variety of sources. Check the county surveyor's office, local aerial photography companies, and universities in the area you're modeling. Perry Squier, who models the Pittsburg, Shawmut & Northern in 1923, found photos of its right-of-way at pennpilot.psu.edu.

Also check Google Earth (earth. google.com; you'll need to download the free software that allows you to zoom, tilt, and rotate the images) for aerial photos – enter an address and it takes you to that location. Or try Microsoft's Virtual Earth or www.local. live.com. The images are considerably sharper than those once offered on the Terraserver site.

There are many online sources of vintage photographs. For example, Virginia Tech's Norfolk & Western Historical Photograph Collection can be accessed at imagebase.lib.vt.edu/browse. php?folio.ID=/trans/nss.

Work now, save later

To be sure, it takes time to pore over maps looking for a stretch of full-size railroad to model or a place where a freelanced railroad might fit into the continental rail network. But it can save time later.

Remember that photo on page 5: It's easier to erase bad ideas from a piece of paper than it is to pry them off the railroad.